Prepared in cooperation with the Triangle J Council of Governments Cape Fear River Flow Study Committee and the North Carolina Department of Environment and Natural Resources, Division of Water Resources

Determination of Flow Losses in the Cape Fear River between B. Everett Jordan Lake and Lillington, North Carolina, 2008–2010

Scientific Investigations Report 2012–5226

U.S. Department of the Interior
U.S. Geological Survey

Determination of Flow Losses in the Cape Fear River between B. Everett Jordan Lake and Lillington, North Carolina, 2008–2010

By J. Curtis Weaver and Kristen Bukowski McSwain

Prepared in cooperation with the Triangle J Council of Governments Cape Fear River Flow Study Committee and the North Carolina Department of Environment and Natural Resources, Division of Water Resources

Scientific Investigations Report 2012–5226

U.S. Department of the Interior
U.S. Geological Survey

U.S. Department of the Interior
KEN SALAZAR, Secretary

U.S. Geological Survey
Marcia K. McNutt, Director

U.S. Geological Survey, Reston, Virginia: 2013

For more information on the USGS—the Federal source for science about the Earth, its natural and living resources, natural hazards, and the environment, visit http://www.usgs.gov or call 1–888–ASK–USGS.

For an overview of USGS information products, including maps, imagery, and publications, visit http://www.usgs.gov/pubprod

To order this and other USGS information products, visit http://store.usgs.gov

Suggested citation:
Weaver, J.C., and McSwain, K.B., 2013, Determination of flow losses in the Cape Fear River between B. Everett Jordan Lake and Lillington, North Carolina, 2008–2010: U.S. Geological Survey Scientific Investigations Report 2012–5226, 76 p., http://pubs.usgs.gov/sir/2012/5226.

Acknowledgments

Support for this study was provided by the North Carolina Division of Water Resources and with the following local cooperators who provided funding through the Triangle J Council of Governments Cape Fear River Flow Study Committee: Cape Fear Public Utility Authority, Town of Cary Water Resources, Chatham County Public Works, City of Durham Water Management, Public Works Commission of the City of Fayetteville, Harnett County Public Utilities, Town of Holly Springs Engineering, Lower Cape Fear Water and Sewer Authority, Town of Morrisville Engineering, Orange Water and Sewer Authority, City of Sanford Public Works, Wake County, and DuPont Company (Fayetteville Works).

The authors acknowledge Mr. Tom Fransen of the N.C. Division of Water Resources; Mr. Mike Schlegel and Mr. Sydney Miller (formerly) of the Triangle J Council of Governments; and Mr. Mick Noland of the Fayetteville Public Works Commission. Their foresight and leadership in areas of water management as well as their assistance and support of this study have improved the understanding of the complex flow patterns in this reach of the Cape Fear River.

The authors gratefully acknowledge the assistance provided by Ms. Ashley Hatchell and Mr. Tony Young of the U.S. Army Corps of Engineers, Wilmington District, during the study. Ms. Hatchell and Mr. Young collaborated with the U.S. Geological Survey on flow releases from Jordan Lake Dam during synoptic discharge measurement runs and other critical data-collection events.

Progress Energy Carolinas provided assistance in this investigation. The authors acknowledge Mr. Mick Greeson and Ms. Robin Bryson for their support in the operation of a streamgage on the diversion canal adjacent to the Cape Fear River. Mr. Greeson and Ms. Bryson were instrumental in providing data concerning water withdrawals from the Cape Fear River into a regional power plant and in obtaining permission to operate a cluster of groundwater piezometers on property owned by the company.

Water-use information provided by Mr. Don Rayno (North Carolina Division of Water Resources), Ms. Carolyn Underwood (Uniboard), Mr. Brad Crawford (Arclin), Mr. Brian Van Gelden (Moncure Plywood), Mr. Charles Powell, Mr. Marty Stewart (Perfomance Fibers), and Mr. Warren Paschal (General Shale Brick) allowed for a better understanding of the industrial water usage in the study area. Information concerning municipal water use from the Cape Fear River was provided by Mr. Victor Czar and Mr. Scott Christiansen (City of Sanford), and Mr. Charles Fiero and Mr. Allan O'Briant (Harnett County Public Utilities).

The authors also acknowledge Mr. William Brooks, owner of the Lockville Hydropower Dam, for providing information concerning the operations at this facility. Access to the Cape Fear River was made possible by the use of privately owned boat ramps graciously provided by Mr. Charles Stirewalt and Mr. Mike Taylor. Additionally, the authors thank Mr. Mike Taylor for allowing property access to monitor a cluster of groundwater piezometers.

Contents

Figures

Tables

Conversion Factors

Inch/Pound to SI

Multiply	By	To obtain
Length		
inch (in.)	2.54	centimeter (cm)
inch (in.)	25.4	millimeter (mm)
foot (ft)	0.3048	meter (m)
mile (mi)	1.609	kilometer (km)
yard (yd)	0.9144	meter (m)
Area		
acre	4,047	square meter (m^2)
square mile (mi^2)	259.0	hectare (ha)
square mile (mi^2)	2.590	square kilometer (km^2)
Flow rate		
foot per second (ft/s)	0.3048	meter per second (m/s)
cubic foot per second (ft^3/s)	0.02832	cubic meter per second (m^3/s)
million gallons per day (Mgal/d)	0.04381	cubic meter per second (m^3/s)
inch per year (in/yr)	25.4	millimeter per year (mm/yr)
Hydraulic gradient		
foot per mile (ft/mi)	0.1894	meter per kilometer (m/km)

Temperature in degrees Celsius (°C) may be converted to degrees Fahrenheit (°F) as follows:

°F=(1.8×°C)+32

Temperature in degrees Fahrenheit (°F) may be converted to degrees Celsius (°C) as follows:

°C=(°F-32)/1.8

Vertical coordinate information is referenced to the National Geodetic Vertical Datum of 1929 (NGVD 29) and North American Vertical Datum of 1988 (NAVD 88) unless otherwise noted.

Horizontal coordinate information is referenced to North American Datum of 1983 (NAD 83).

Altitude, as used in this report, refers to distance of groundwater level (or water table) above the vertical datum.

Elevation, as used in this report, refers to distance of surface-water level (stage or gage height) or land surface above the vertical datum.

Abbreviations

DMR	discharge monitoring report
DTS	distributed temperature sensing (survey)
EM	electromagnetic
ET	evapotranspiration
FAO	Food and Agricultural Organization
FERC	Federal Energy Regulatory Commission
FO-DTS	fiber-optic distributed-temperature sensing
GPS	global positioning system
Hz	hertz
kW	kilowatt
NOAA	National Oceanic and Atmospheric Administration
NWIS	National Water Information System
ppm	parts per million
PVC	polyvinyl chloride
PZ	piezometer (groundwater transect)
RTK	Real-Time Kinematic (GPS equipment)
TIR	thermal infrared (imaging)
USACE	U.S. Army Corps of Engineers
USGS	U.S. Geological Survey

Determination of Flow Losses in the Cape Fear River between B. Everett Jordan Lake and Lillington, North Carolina, 2008–2010

By J. Curtis Weaver and Kristen Bukowski McSwain

Abstract

During 2008–2010, the U.S. Geological Survey conducted a hydrologic investigation in cooperation with the Triangle J Council of Governments Cape Fear River Flow Study Committee and the North Carolina Division of Water Resources to collect hydrologic data in the Cape Fear River between B. Everett Jordan Lake and Lillington in central North Carolina to help determine if suspected flow losses occur in the reach. Flow loss analyses were completed by summing the daily flow releases at Jordan Lake Dam with the daily discharges at Deep River at Moncure and Buckhorn Creek near Corinth, then subtracting these values from the daily discharges at Cape Fear River at Lillington. Examination of long-term records revealed that during 10,227 days of the 1983–2010 water years,[1] 408 days (4.0 percent) had flow loss when conditions were relatively steady with respect to the previous day's records. The flow loss that occurred on these 408 days ranged from 0.49 to 2,150 cubic feet per second with a median flow loss of 37.2 cubic feet per second. The months with the highest number of days with flow losses were June (16.7 percent), September (16.9 percent), and October (19.4 percent).

A series of synoptic discharge measurements made on six separate days in 2009 provided "snapshots" of overall flow conditions along the study reach. The largest water diversion is just downstream from the confluence of the Haw and Deep Rivers, and discharges substantially decrease in the main stem downstream from the intake point. Downstream from Buckhorn Dam, minimal gain or loss between the dam and Raven Rock State Park was noted.

Analyses of discharge measurements and ratings for two streamgages—one at Deep River at Moncure and the other at Cape Fear River at Lillington—were completed to address the accuracy of the relation between stage and discharge at these sites. The ratings analyses did not indicate a particular time during the 1982–2011 water years in which a consistent bias occurred in the computations of discharge records that would indicate false flow losses.

A total of 34 measured discharges at a streamgage on the Haw River below B. Everett Jordan Lake near Moncure were compared with the reported hourly flow releases from Jordan Lake Dam. Because 28 of 34 measurements were within plus or minus 10 percent of the hourly flow releases reported by the U.S Army Corps of Engineers, use of the current discharge computation tables for reporting Jordan Lake Dam flow releases is generally supported.

A stage gage was operated on the Cape Fear River at Buckhorn Dam near Corinth to collect continuous stage-only records. Throughout the study period, flow over the dam was observed along its length, and flow loss within the study reach is not attributed to river-level fluctuations at the dam.

Water-use information and (or) data were obtained for five industrial facilities, a regional power utility, two municipalities, one small hydropower facility on the Deep River, and one quarry operation also adjacent to the Deep River. The largest water users are the regional power producer, a small hydropower operation, and the two municipalities. The total water-use diversions for these facilities range from almost 25.5 to 38.5 cubic feet per second (39.5 to 59.5 million gallons per day) during the winter and summer periods, respectively. This range is equivalent to 69 to 104 percent of the 37 cubic feet per second median flow loss.

The Lockville hydropower station is on the Deep River about 1 mile downstream from the streamgage near Moncure. Run-of-river operations at the facility do not appear to affect flow losses in the study reach. The largest water user in the study area is a regional power producer at a coal-fired power-generation plant located immediately adjacent to the Cape Fear River just downstream from the confluence of the Haw and Deep Rivers. Comparisons of daily water withdrawals, supplied by the regional power producer, and discharge records at a streamgage on the diversion canal indicated many days when consumption exceeded the producer's estimates for the cooling towers. Uncertainty surrounding reasonable estimates of consumption remained in effect at the end of the study.

Data concerning evaporative losses were compiled using two approaches—an analysis of available pan-evaporation data from a National Weather Service cooperative observer station in Chapel Hill, North Carolina; and a compilation

---

[1] A water year is the 12-month period October 1 through September 30, designated by the calendar year in which the period ends. For example, the 2010 water year began on October 1, 2009, and ended on September 30, 2010.

of reference open-water evaporation computed by the State Climate Office of North Carolina. The potential flow loss by evaporation from the main stem and the Deep River was estimated to be in the range of 4 to 14 cubic feet per second during May through October, equivalent to 10 to 38 percent of the 37 cubic feet per second median flow loss.

Daily water-use diversions and evaporation losses were compared to flow-loss occurrences during the period April 2008 through September 2010. In comparing the surface-water, water-use, and evaporation data compiled for 2008–2010, it is evident that documented water diversions combined with flow losses by open-water evaporation can exceed the net flow gains in the study area and result in flow losses from the reach.

Analysis of data from a streamgage downstream from the regional power plant on the diversion canal adjacent to the Cape Fear River provided insight into the occurrence of an apparent flow loss at the streamgage at Lillington. Assessment of the daily discharges and subsequent hydrographs for the canal streamgage indicated at least 24 instances during the study when the flows suddenly changed by magnitudes of 100 to more than 200 cubic feet per second, resulting in a noted time-lag effect on the downstream discharges at the Lillington streamgage, beginning 8 to 16 hours after the sudden flow change.

A fiber-optic distributed temperature-sensing survey was conducted on the Cape Fear River at the Raven Rock State Park reach August 12–14, 2009, to determine if the presence of diabase dikes were preferentially directing groundwater discharge. No temperature anomalies of colder water were measured during the survey, which indicated that at the time of the survey that particular reach of the Cape Fear River was a "no-flow" or losing stream.

An aerial thermal-infrared survey was conducted on the Haw and Cape Fear Rivers on February 27, 2010, from Jordan Lake Dam to Lillington to qualitatively delineate areas of groundwater discharge on the basis of the contrast between warm groundwater discharge and cold surface-water temperatures. Discharge generally was noted as diffuse seepage, but in a few cases springs were detected as inflow at a discrete point of discharge.

Two reaches of the Cape Fear River (regional power plant and Bradley Road reaches) were selected for groundwater monitoring with a transect of piezometers installed within the flood plain. Groundwater-level altitudes at these reaches were analyzed for 1 water year (October 1, 2009, to September 30, 2010). Data collected as part of this study represent only a brief period of time and may not represent all conditions and all years; however, the data indicate that, during the dry summer months, the Cape Fear River within the study area is losing an undetermined quantity of water through seepage.

Analyses completed during this investigation indicate a study reach with complex flow patterns affected by numerous concurrent factors resulting in flow losses. The causes of flow loss could not be solely attributed to any one factor. Among the factors considered, the occurrences of water diversions and evaporative losses were determined to be sufficient on some days (particularly during the base-flow period) to exceed the net gain in flows between the upstream and downstream

ends of the study area. Losses by diversions and evaporation can exceed the median flow loss of 37 cubic feet per second, which indicates that flow loss from the study reach is real. Groundwater data collected during 2009–2010 indicate the possibility of localized flow loss during the summer, particularly in the impounded reach above Buckhorn Dam. However, no indication of unusual patterns was noted that would cause substantial flow loss by groundwater and surface-water interaction at the river bottom.

Introduction

Streamflows on large rivers are affected by a number of factors, including precipitation, evaporative losses, and underlying geologic and soils characteristics within a basin or along a given reach. Other factors include diversions to and from the rivers for water use as well as the impoundment of water behind structures (both actively regulated and run-of-river structures) that alter the normal regime of streamflow. The combined effects of these factors can result in complex flow patterns that indicate apparent or real flow losses for a reach of interest. One such reach suspected of being a losing reach is the Haw River and Cape Fear River between B. Everett Jordan Lake Dam (hereafter referred to as Jordan Lake Dam) and Lillington in Harnett County in central North Carolina (fig. 1). The Cape Fear River is formed at the confluence of the Haw and Deep Rivers. Real flow losses would reflect consumptive use somewhere within the intervening drainage area or the presence of a losing reach attributed to natural factors (underlying geology). Apparent flow losses would reflect flow dynamics that occur primarily as a result of water diversion through canals, storage, or potential flow regulations at the dams, and (or) measurement error at observation points.

In accordance with the U.S. Army Corps of Engineers' (USACE) water-control plan, water is released from Jordan Lake Dam (fig. 1, site D1) to maintain a normal minimum target flow of 600 cubic feet per second (ft^3/s), with a tolerance of plus or minus (+/−) 50 ft^3/s, at the U.S. Geological Survey (USGS) streamgaging station (or streamgage) on the Cape Fear River at Lillington (fig. 1, site 28) in Harnett County (Michael A. (Tony) Young, U.S. Army Corps of Engineers, written commun., December 20, 2011). The USACE estimates flow releases at Jordan Lake Dam based on a rating between the service gate openings in the intake tower and theoretically computed flows through the openings based on the lake level.

During some low-flow conditions, the sum of the estimated releases from Jordan Lake Dam and the observed streamflow at a second long-term streamgage on the Deep River at Moncure (fig. 1, site 3) in Chatham County is greater than the observed streamflow downstream at the streamgage at Lillington. In addition, flows at the Lillington streamgage have been observed to decrease by as much as 50 to 150 ft^3/s over a period of 12 to 24 hours for unknown reasons. Consequently, either the

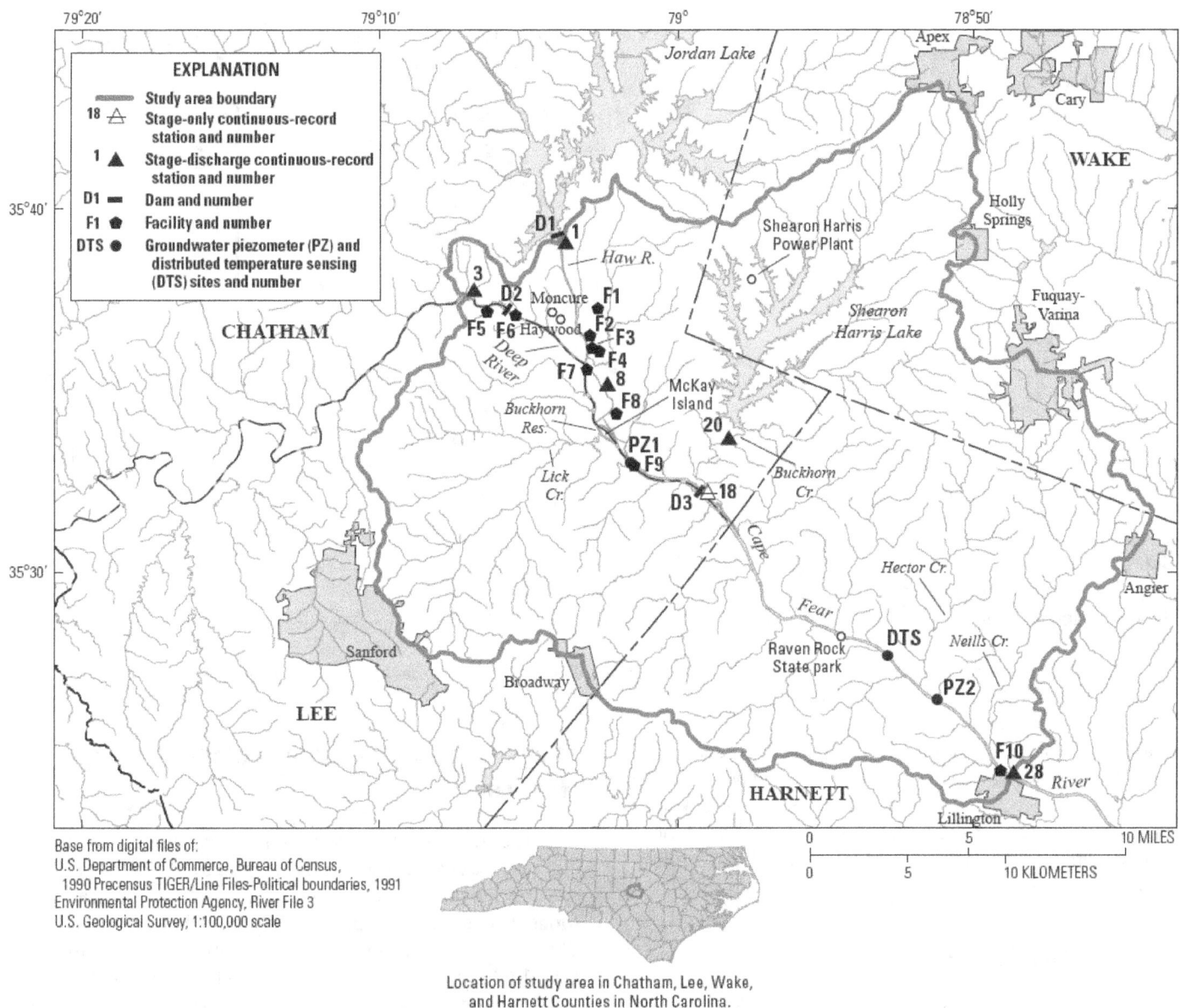

Figure 1.   Location of study area between B. Everett Jordan Lake and Lillington, North Carolina.

target flow at Lillington is not met, or additional water must be released from Jordan Lake in an attempt to meet the target.

Repeated occurrences of suspected flow loss between Jordan Lake Dam and Lillington have raised questions about the factors that affect flows (or discharges) in this reach, whether or not flow losses are real, and the source(s) behind the flow losses. The difference between the inflow (sum of releases from Jordan Lake Dam and flow in the Deep River at Moncure) and outflow from the reach (flow in the Cape Fear River at Lillington) could be because either (1) flow is measured inaccurately at one or more of the three measurement locations; or (2) flow is lost from the reach as a result of surface-water diversions, possible alterations because of impoundment behind dam structures, groundwater withdrawals, evaporative losses, recharge of the groundwater system, or some combination of these factors.

During 2008–2010, the USGS conducted a hydrologic investigation in cooperation with the Triangle J Council of Governments Cape Fear River Flow Study Committee and the North Carolina Division of Water Resources to characterize natural and manmade factors that affect the study reach and to collect hydrologic data to help determine if the losses indicated in the initial analysis of data are real or apparent.

This study specifically addresses the USGS science strategy goal, "A Water Census of the United States: quantifying, forecasting, and securing freshwater for America's Future" (U.S. Geological Survey, 2007a). The study also addresses the water census priority issue of the USGS 2011 Federal-State Water Cooperative Program, which helps to meet the goal of better understanding water availability and use (U.S. Geological Survey, 2010a). Finally, the study meets the science plan goal of the USGS North Carolina Water Science Center

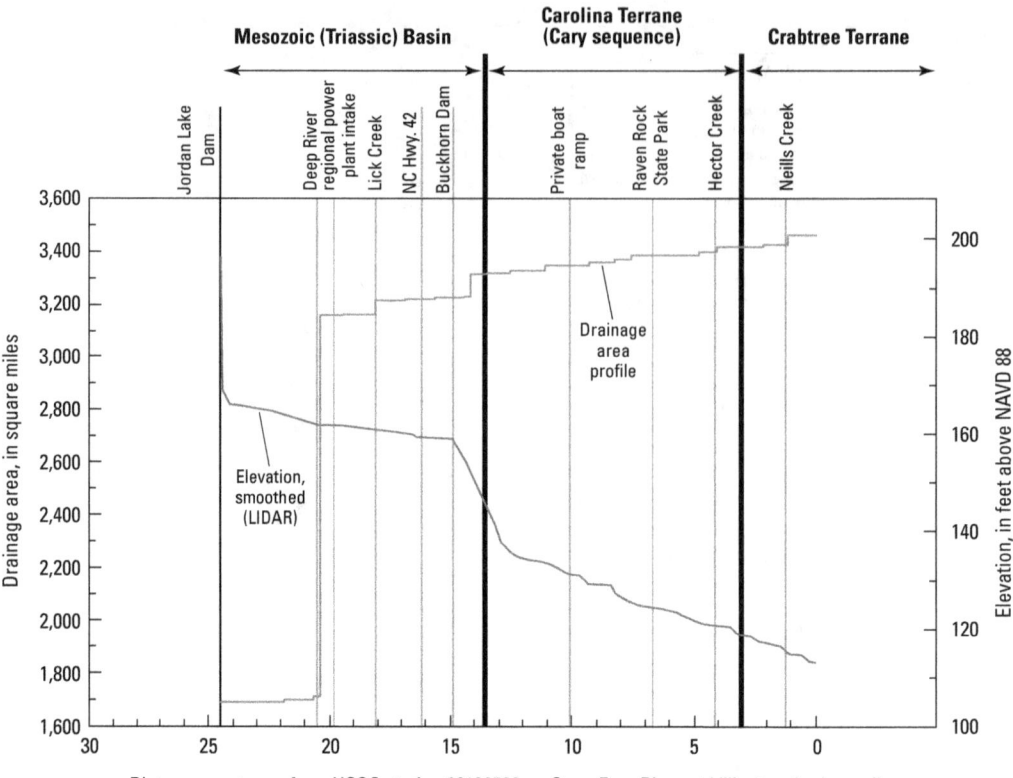

Figure 2. Drainage area and elevation profiles on the Haw and Cape Fear Rivers between Jordan Lake Dam and Lillington, North Carolina.

(Bales and others, 2004) to understand increased demands on water resources through awareness of State and local water use, greatly enhancing the water-use database with an understanding of the regional, environmental, climatic, and economic factors affecting water use in North Carolina streams and rivers.

Purpose and Scope

The purpose of this report is to describe the causes of flow losses in the Cape Fear River between Jordan Lake Dam and Lillington, North Carolina. The report presents varied data-collection and data-analysis approaches selected to contribute toward an improved understanding of the possible causes of the flow losses. Results of the various approaches are interpreted to develop a conceptual model of the man-made or natural causes of the flow losses based on the new data collection and analysis.

The scope of the report includes results of flow analyses of long-term records for the 1983–2010 water years when losses were determined, description and results of the synoptic discharge measurements that were collected during 2009, assessments of water-use data compiled during the study, estimated losses by evaporation, and discussion concerning the groundwater data collected at two piezometer well clusters during 2009–2010. Results of a 2009 magnetometer survey and an aerial infrared flyover completed in February 2010 also are presented.

Study Area and Possible Factors Affecting Flow Loss

The study area for this investigation is the intervening drainage area between Jordan Lake Dam near Moncure (fig. 1, site D1) and the USGS streamgages on the Deep River at Moncure (site 3) and Cape Fear River at Lillington (site 28). The drainage area along this 24.5-mile (mi) reach increases from 1,689 square miles (mi^2) at Jordan Lake Dam to 3,464 mi^2 at the Lillington streamgage (resulting in a 1,775-mi^2 difference in drainage area). The intervening drainage area in this reach includes the portion of the Deep River basin downstream from the Moncure streamgage (site 3), which has a drainage area of 1,434 mi^2. Accounting for this drainage area results in an inter-vening drainage area of 341 mi^2 for the study area.

As discussed in a subsequent section, the final study area did not include the basin upstream from a long-term USGS streamgage on Buckhorn Creek near Corinth (fig. 1, site 20). Buckhorn Creek is the largest tributary in the intervening drainage area. By not including this basin, the final intervening (ungaged) drainage area for this study is nearly 265 mi^2.

Upstream from Buckhorn Dam (site D3), the Cape Fear River is characterized by fairly level water surfaces reflective of the lake setting. The average slope of the water surface between Jordan Lake Dam and Buckhorn Dam is approximately 0.5 foot per mile (ft/mi; fig. 2). Water depths range from 5–10 feet (ft) downstream from Jordan Lake Dam to 20–25 ft just upstream from Buckhorn Dam. Downstream from Buckhorn Dam, the river bottom is characterized by rock outcrops of varying size, ranging from small cobbles to substantial boulders. Navigation using motorized craft is limited in this section of the river, particularly during base-flow conditions. Between Buckhorn Dam and the Lillington streamgage (site 28), the channel slope varies from about 7 ft/mi in the 3-mi reach below the dam to about 1.6 ft/mi in the downstream most 11.5-mi reach (fig. 2).

A diversion canal runs adjacent to the Cape Fear River just downstream from the confluence of the Haw and Deep Rivers to immediately upstream from Buckhorn Dam. The area of the basin drained by the canal is approximately 10.3 mi². Water used in the regional power plant for cooling purposes travels the length of the canal about 6 mi before merging back into the Cape Fear River at the dam.

Streamflow records are available for five locations in the study area (table 1). Streamflow records were collected on the Deep River at Moncure (fig. 1, site 3) beginning in July 1930, and records were collected on the Cape Fear River at Lillington (site 28) beginning in December 1923. Prior to October 1992, streamflow records also were collected on the Haw River below B. Everett Jordan Dam near Moncure (site 1) and were published as Haw River near Haywood (USGS station 02098200) during the 1966–78 water years before and during the construction of Jordan Lake just downstream from the dam. Since April 2008, continuous streamflow records have been collected at a streamgaging station (site 8) downstream from a power plant on a diversion canal adjacent to the Cape Fear River. Continuous-record discharge also is measured on Buckhorn Creek (site 20) downstream from Shearon Harris Lake, which is owned and operated by the regional power producer to supply cooling water to the Shearon Harris Power Plant (fig. 1) in southwest Wake County. No active or inactive continuous groundwater monitoring sites are located within the study area.

Consideration of possible flow losses is complicated by several factors that may affect the overall flow patterns in the study reach. One factor is two low-head, run-of-river dams located in the study area. The Lockville Hydropower Dam (fig. 1, site D2) is on the Deep River about 2.7 mi upstream from the mouth of the river. The dam is used for hydropower generation at a power station (site F6) about 0.5 mi downstream from the dam. Ownership of the dam changed in 2003, and power production has occurred at this site since early 2008. Further discussion about this dam and the operations are provided in a subsequent section.

Buckhorn Dam (figs. 1 (site D3), 3) is on the Cape Fear River, 13.8 mi upstream from the U.S. Highway 401 bridge and 5.9 mi downstream from the confluence of the Haw and Deep Rivers. The dam, which is approximately 1,100 ft in length between the edges of the river, was completed and the reservoir was filled in 1908 (Ragland and others, 2003). This dam was used until December 1962 by a regional power producer to regulate flows for power production. The estimated surface area of the river upstream from the dam to Jordan Lake Dam and the Lockville Dam is about 460 acres. The volume of water in storage behind Buckhorn Dam is reported to be 69.7 million ft³ (Walters and others, 2006). The dam has been a run-of-river

structure with no active regulation of downstream flows since December 1962 and continues to provide water storage for cooling purposes at the power plant.

Another factor that may affect possible flow losses is water withdrawals and return point-source discharges to and from the Haw and Cape Fear Rivers (or main stem) in the study area. The largest withdrawal is associated with the operation of the coal-fired power plant (fig. 1, site F7) owned by a regional power producer located adjacent to the Cape Fear River just downstream from the confluence of the Deep and Haw Rivers. Water is withdrawn from the river into the plant and discharged to the diversion canal that runs generally parallel to the main stem, merging back into the river immediately upstream from Buckhorn Dam (site D3). The average daily withdrawal and return discharge in 1998 for the plant was about 207 and 204 million gallons per day (Mgal/d), respectively (Weaver, 2001). As previously noted, a streamgaging station (site 8) was installed on the canal in March 2008 to provide additional information about the flow through this facility. No major water withdrawals and return point-source discharges were identified or observed to or from the Deep River during the study period.

Two municipal withdrawals for water-supply purposes also occur in the study area. The first withdrawal is on the Cape Fear River upstream from N.C. Highway 42 (fig. 1, site F9), and the associated wastewater discharge is on Buffalo Creek, a tributary to the Deep River upstream from the study area. The second municipal withdrawal (site F10) is on the Cape Fear River upstream from the streamgage at Lillington (site 28), and the associated wastewater discharge is downstream from the streamgage and the study area. Because the associated wastewater discharges for these two withdrawals are located outside of the study area, the withdrawals for these two intakes were considered equivalent to 100 percent consumptive use within the study area.

Water withdrawals and return discharges also are made by five private industries (fig. 1), four of which (sites F1–F4) are located on the Haw River upstream from its confluence with the Deep River. The fifth industry (site F8) withdraws water from the diversion canal adjacent to the Cape Fear River. Groundwater withdrawals by a quarry operation (site F5) upstream from the Lockville hydropower dam also were identified as a possible cause of flow loss, because groundwater is intercepted that would be otherwise discharged to the Deep River.

Geologic Setting

The geologic setting is complex in the study area because it is located near the Fall Line between the Piedmont and Coastal Plain Physiographic Provinces. The majority of the study area is in the Piedmont and includes the Mesozoic basin (44 percent) and members of the Carolina Terrane (Cary Sequence, 32.1 percent; Virgilina Sequence, 2 percent) and Crabtree Terrane (4.4 percent, fig. 4; Hibbard and others, 2002). Coastal Plain sediments (17.5 percent) are present as a thin veneer in the southeastern part of the study area (fig. 4;

Table 1. Continuous- and partial-record U.S. Geological Survey surface-water sites in the Cape Fear River study area, North Carolina.

[USGS, U.S. Geological Survey; °, degree; ', minute; ", second; mi², square mile; CR, continuous-record gaging station; PR, partial-record synoptic site; SO, stage only]

Site index number (figs. 1, 6, 7)	USGS station number	Station name	Latitude	Longitude	Drainage area, mi²	Site type	Period of record	Remarks
1	02098198	Haw R below B. Everett Jordan Dam near Moncure, NC	35°39'01"	79°03'58"	1,689	CR	October 1965 to current	Prior to October 1992, discharge records published; prior to September 1978, published as "Haw River near Haywood, NC" (USGS 02098200, located immediately downstream from current location of 02098198).
2	02098210	Haw River above Deep River near Haywood, NC	35°36'00"	79°03'03"	1,708	PR		
3	02102000	Deep River at Moncure, NC	35°37'38"	79°06'57"	1,434	CR	July 1930 to current	Diurnal fluctuation and some regulation at low flow caused by small power plants upstream from station.
4	0210204915	Deep River above mouth near Moncure, NC	35°36'22"	79°03'57"	1,448	PR		
5	02102050	Cape Fear River above powerplant intake near Moncure, NC	35°35'40"	79°03'08"	3,157	PR		
6	02102082	Cape Fear River below powerplant intake near Moncure, NC	35°35'03"	79°03'05"	3,157	PR		
7	02102090	Cape Fear River above railroad bridge near Rosser, NC	35°34'11"	79°02'48"	3,166	PR		
8	02102094	Cape Fear powerplant discharge canal near Brickhaven, NC	35°35'03"	79°02'28"	0.22	CR	April 2008 to current	
9	0210209450	Cape Fear powerplant discharge canal above juncture at Brickhaven, NC	35°34'21"	79°02'26"	1.32	PR		
10	0210209475	Cape Fear powerplant discharge canal below dam at Brickhaven, NC	35°34'17"	79°02'32"	1.33	PR		
11	0210209625	Unnamed stream between canal and Cape Fear River at Brickhaven, NC	35°34'15"	79°02'30"	1.33	PR		
12	0210209650	Cape Fear powerplant discharge canal below juncture at Brickhaven, NC	35°34'12"	79°02'26"	1.00	PR		
13	02102159	Lick Creek near Rosser, NC	35°33'33"	79°03'16"	45.6	PR		
14	0210215960	Cape Fear River above NC Highway 42 near Corinth, NC	35°33'02"	79°01'36"	3,226	PR		

Table 1. Continuous- and partial-record U.S. Geological Survey surface-water sites in the Cape Fear River study area, North Carolina.—Continued

[USGS, U.S. Geological Survey; °, degree; ′, minute; ″, second; mi², square mile; CR, continuous-record gaging station; PR, partial-record synoptic site; SO, stage only]

Site index number (figs. 1, 6, 7)	USGS station number	Station name	Latitude	Longitude	Drainage area, mi²	Site type	Period of record	Remarks
15	0210215995	Cape Fear powerplant discharge canal at NC Highway 42 near Corinth, NC	35°33′02″	79°01′24″	8.54	PR		
16	02102176	Cape Fear powerplant discharge canal above mouth near Corinth, NC	35°32′36″	78°59′52″	10.3	PR		
17	02102177	Cape Fear River upstream from Buckhorn Dam near Corinth, NC	35°32′27″	78°59′32″	3,234	PR		
18	02102178	Cape Fear River at Buckhorn Dam near Corinth, NC	35°32′27″	78°59′21″	3,235	SO	November 2008 to June 2010	
19	0210217810	Cape Fear River downstream from Buckhorn Dam near Corinth, NC	35°32′09″	78°59′04″	3,235	PR		
20	02102192	Buckhorn Creek near Corinth, NC	35°33′35″	78°58′22″	76.3	CR	June 1972 to current	Flow regulated since December 1, 1980, by Shearon Harris Lake
21	02102240	Cape Fear River adjacent to Bay Street near Cokesbury, NC	35°29′13″	78°57′06″	3,354	PR		
22	02102265	Cape Fear River at Raven Rock State Park group camp, NC	35°27′49″	78°52′59″	3,391	PR		
23	02102278	Cape Fear River above Hector Creek near Chalybeate, NC	35°26′53″	78°51′54″	3,400	PR		
24	02102280	Hector Creek near Chalybeate, NC	35°28′00″	78°51′29″	17.4	PR		
25	02102283	Cape Fear R adjacent to Bradley Road near Lillington, NC	35°26′28″	78°51′16″	3,420	PR		
26	02102289	Cape Fear River above Neills Creek near Lillington, NC	35°25′22″	78°49′38″	3,425	PR		
27	02102480	Neills Creek near Lillington, NC	35°25′42″	78°49′27″	37.6	PR		
28	02102500	Cape Fear River at Lillington, NC	35°24′22″	78°48′47″	3,464	CR	December 1923 to current	Flow regulated since September 1981 by B. Everett Jordan Lake

Figure 3.   Cape Fear River at Buckhorn Dam (site D3) near Corinth, North Carolina. (Photo by J. Curtis Weaver on April 27, 2009)

Hibbard and others, 2002). Coastal Plain sediments that once covered the entire southeastern part of the study area have eroded as sea level declined and the Cape Fear River incised into the underlying Piedmont rocks.

Questions have been raised as to whether the underlying geologic setting could be a factor in flow losses in the study area. The presence of diabase dikes that intersect the river could act as impermeable boundaries to groundwater discharge (McSwain, 2009). Diabase dikes are high-angle, vertically intruded geologic features of mafic igneous rock with less thickness than length. The thickness of diabase dikes can vary from inches to tens of feet, and length can exceed many miles. As a result of weathering over time, diabase dikes become exposed at the surface (fig. 5) and can intersect river bottoms.

Diabase dikes were mapped previously by Burt and others (1978) in rocks of the Mesozoic basin and Carolina Terrane. As a result of rifting activity, isolated tabular diabase dikes may extend into the Mesozoic basin sedimentary rocks. Weaver and Pope (2001) noted that rocks and soils of the Mesozoic basin generally have low permeabilities, and perched water tables are common. Additionally, they noted that soils in the Mesozoic basin support a lower potential for sustained base flows, and they identified a number of streamflow sites in the Mesozoic basin that were determined to have 7-day, 10-year low-flow discharges equal to zero flow. In the reach downstream from Buckhorn Dam (figs. 1, 4), the Cary sequence of the Carolina Terrane contains high-grade metamorphic gneiss and schist intruded by granite plutons and diabase dikes (Burt and others, 1978; Hibbard and others, 2002). Burt and others (1978) did not locate diabase dikes within the metamorphic slates of Crabtree Terrane, metamorphic volcanic and sedimentary rock of the Carolina Terrane, Virgilina sequence, or Coastal Plain sediments of the study area.

Data Collection and Methods for Assessing Flow Patterns in the Cape Fear River

Assessment of the flow patterns in the study reach required a multidiscipline approach. In addition to existing surface-water records, additional surface- and groundwater data were collected during the study to enable further analyses of flow into and out of the study area. Data also were collected regarding water use in the study area and the potential evaporative losses from the study reach. Elements of the study also included a geophysical survey, a distributed temperature-sensing survey, and an aerial infrared survey to help define the groundwater and surface-water interaction along the study reach.

Surface Water

The collection of surface-water data involved the (1) operation of currently active streamgaging stations in the study area along with the reactivation of a site on the Haw River below Jordan Lake Dam, (2) activation of a stage-only streamgage on the Cape Fear River at Buckhorn Dam, (3) completion of synoptic discharge measurements along the main stem and selected tributaries, (4) estimation of flow losses by evaporation, and (5) compilation of available water-use information for industrial and municipal facilities adjacent to the Haw and Cape Fear Rivers in the study area. Evaluations of historical discharge records were completed to document the occurrence and magnitudes of suspected flow losses. Analyses of ratings at the two long-term streamgaging stations also were completed to determine if the computation of discharge records could result in falsely detected flow loss.

Flow Loss Analyses

Flow analyses were completed using historical USGS discharge records and flow releases reported by the USACE from Jordan Lake Dam to further characterize the magnitude and frequency of flow loss occurrences between the dam and the streamgage on the Cape Fear River at Lillington. Both instantaneous and daily discharge records available for the 1983–2010 water years were compiled. When comparing flows at upstream and downstream locations, the upstream flow can be higher than the downstream flow due to a peak flow traveling along the reach. The time of travel and passage of peaks between upstream and downstream locations limit the usefulness of a comparison between simultaneous instantaneous discharges to assess the occurrence of flow losses. Likewise, using average flows over a large number of days (for example, 14- or 28-day average flows) also may result in the loss of detailed fluctuations in the comparisons. Therefore, both daily mean discharge records and 7-day average flows were used as the primary

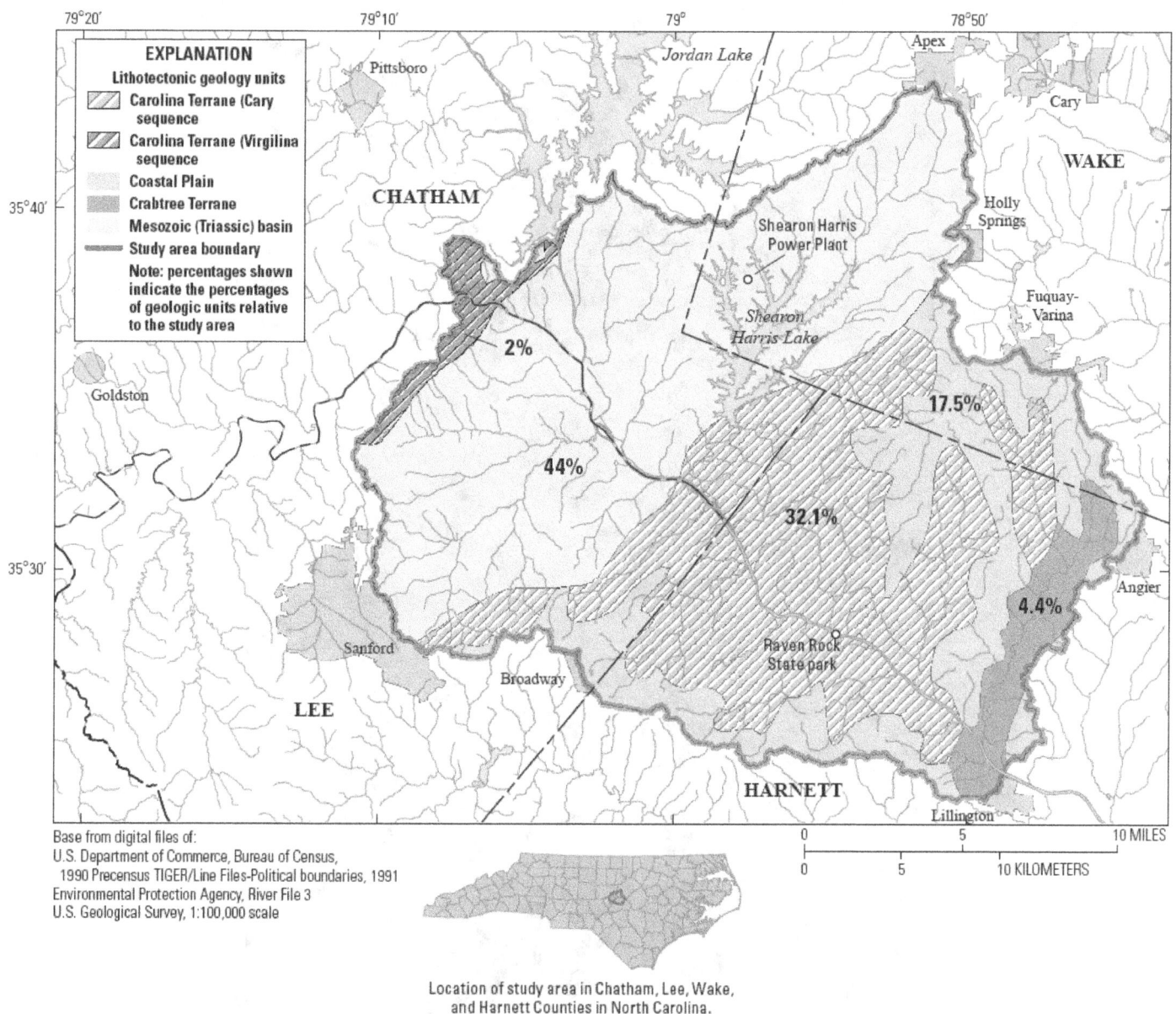

Figure 4. Lithotectonic geologic units in the Cape Fear River study area, North Carolina (from Hibbard and others, 2006).

dataset for further analysis to identify days on which flow loss occurred and the magnitude of such losses.

Synoptic Discharge Measurements

A series of six synoptic discharge measurements were made during 2009 along the Haw and Cape Fear Rivers and selected tributaries to create a "snapshot" of the flow variations along the main stem. Completing the synoptic measurements helped to determine if similar flow variations among the series could identify a particular point along the main stem where flow losses were consistently evident.

The synoptic discharge measurements were made on April 27, June 25, July 22, August 18, September 9, and October 1,

2009. Tributaries were selected based on the amount of change in an intervening drainage area contributed by the tributary basin. For example, a discharge measurement was made on a tributary if the intervening drainage area increased by 5 percent or more. The following tributaries were selected: Deep River, Lick Creek, Buckhorn Creek, Hector Creek, and Neills Creek (fig. 1).

During the April 27 synoptic series, measurements were made at 20 locations in the study area, 13 of which were along the main stem. The additional 7 measurements were made on the adjacent diversion canal and the selected tributaries (2 locations on the Deep River and 1 each on the four smaller tributaries). Post-synoptic processing of the April 27 measurements indicated a breach between the diversion canal and main stem at McKay Island. Following a detailed reconnaissance of the diversion canal on May 13, 6 new locations were

Figure 5.   An example of a diabase dike (dark feature near center) that vertically intrudes the subsurface of a quarry near Cary, North Carolina. (Photo by Richard Bolich, North Carolina Division of Water Quality, April 5, 2005).

added for inclusion in the synoptic series: 1 location on the main stem and 5 locations on the adjacent diversion canal and unnamed stream (table 1, sites 7, 9, 11, 12, 15, 16) between the diversion canal and main stem that constituted the breach.

During the remaining five synoptic series (June 25, July 22, August 18, September 9, and October 1), measurements were made at 26 locations: 14 on the main stem and 12 on the adjacent diversion canal (6) and the tributaries (6). Among the 26 locations, 17 and 9 sites were located upstream and downstream, respectively, from Buckhorn Dam (figs. 6, 7, respectively). Except for a few instances when equipment or the boat malfunctioned, measurements were made at each location

during the synoptic series. As part of each synoptic series, the USGS collaborated with the Wilmington District USACE to set and maintain steady flow releases from Jordan Lake Dam as the measurements were collected. Efforts were made to complete each synoptic series within the course of a single day.

Velocity and corresponding discharge measurements were made in the water column using acoustic Doppler current profilers and velocimeters (Mueller and Wagner, 2009). Use of such equipment allowed the completion of measurements at all locations within 1 day. However, as discussed further in subsequent sections, the presence of backwater conditions, extremely low velocities, wind effects, and eddies in the main

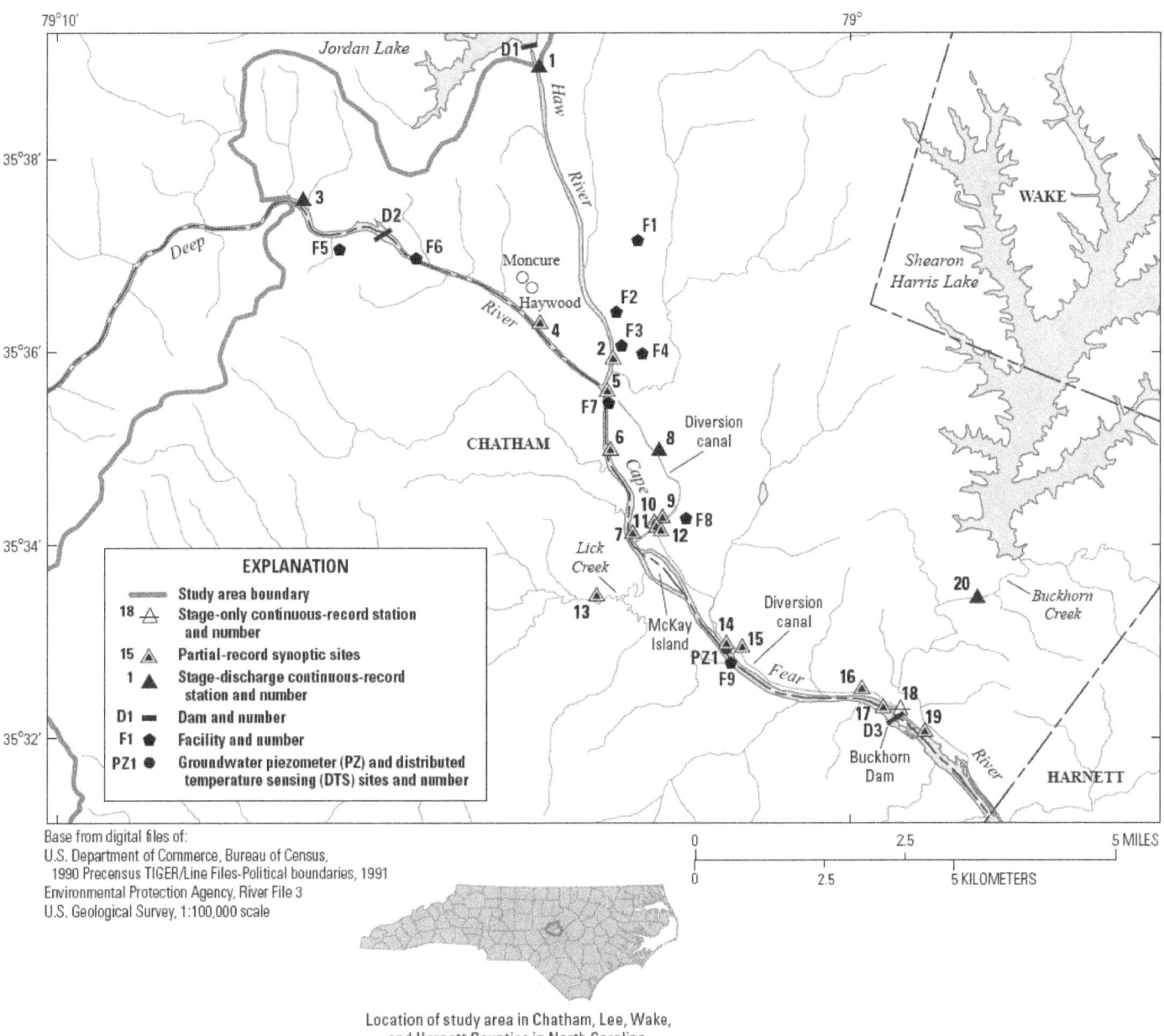

Figure 6. Locations upstream from Cape Fear River at Buckhorn Dam where discharge measurements were made during the series of synoptic measurements in 2009.

stem upstream from Buckhorn Dam affected the quality of the discharge measurements at sites upstream from the dam. Following post-processing and review of the discharge measurements, some measurements were discarded because of poor measuring conditions.

Streamgage Data Collection

At the beginning of the study, four continuous-record streamgages were in operation and were used to provide additional data for assessment in the flow analyses (table 1, sites 3, 8, 20, 28). The collection of continuous discharge data also was reactivated at a streamgage on the Haw River just downstream from Jordan Lake Dam (site D1), which was previously operated from October 1965 to September 1992 (published as USGS station 02098200, Haw River near Haywood prior to October 1978). Streamflow at this site is affected by backwater conditions, the effects of storage behind Buckhorn Dam and (or) inflow from the Deep River into the Cape Fear River. Thus, the streamgage was operated as an index-velocity station where velocity data and stage were collected to develop a rating for computing discharges.

Continuous stage-only records were collected on the Cape Fear River at Buckhorn Dam near Corinth (table 1, site 18) to determine whether water levels fell below the crest of the dam during periods of suspected flow loss. Data

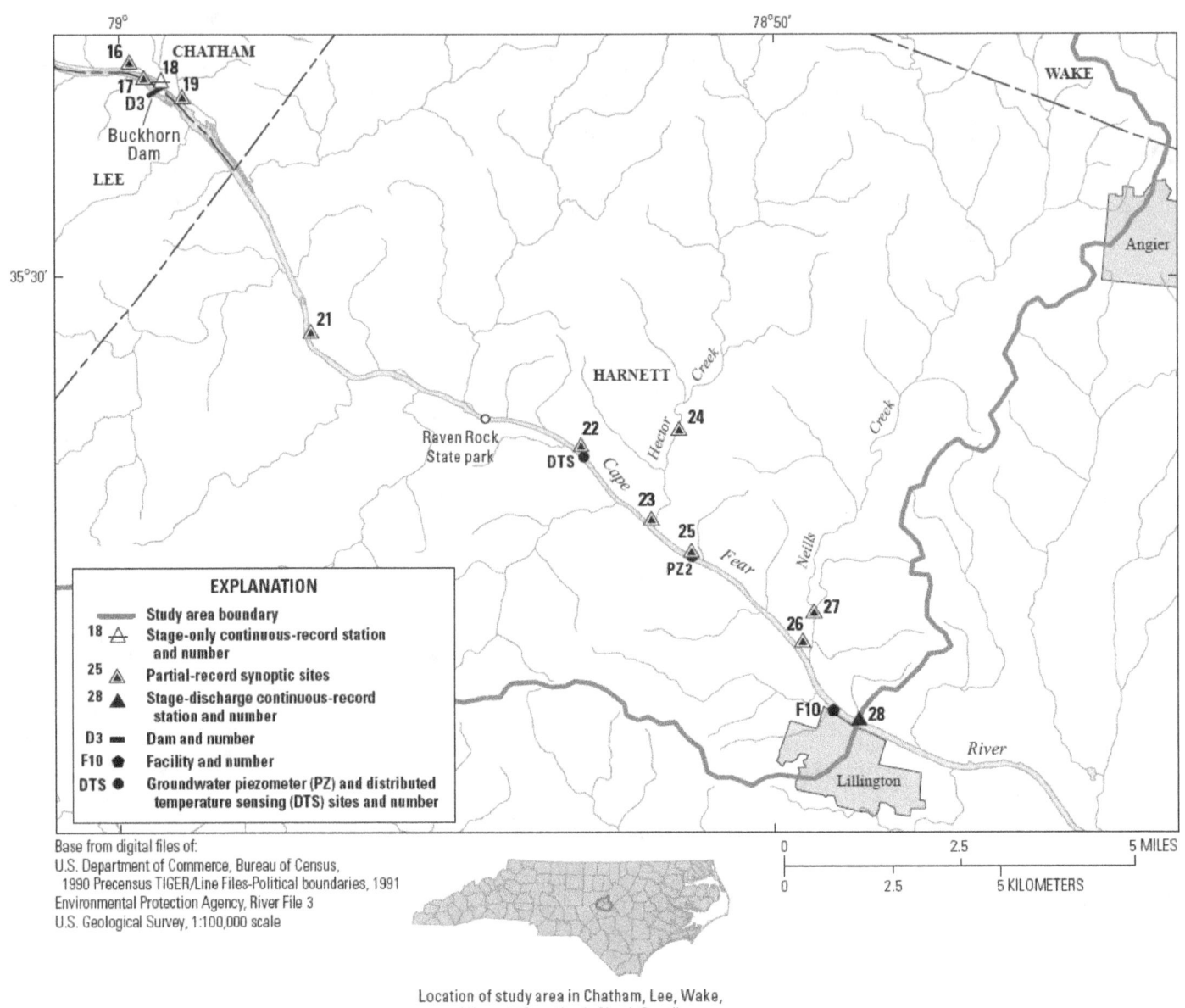

Location of study area in Chatham, Lee, Wake, and Harnett Counties in North Carolina.

Figure 7. Locations downstream from Cape Fear River at Buckhorn Dam where discharge measurements were made during the series of synoptic measurements in 2009.

collection began in November 2008 and continued, except for a brief interruption during January 2008 when recording equipment was stolen, until late June 2010 when the entire streamgage structure and recording equipment were stolen.

Analyses of Ratings and Measured Discharges

Analyses were completed on ratings and measured discharges for the 1982–2011 water years at the two long-term streamgages on the Deep River at Moncure and Cape Fear River at Lillington (figs. 6, 7, sites 3, 28, respectively). The analyses were completed to determine if the computation of discharge records could be a factor in flow loss. In other words, could there be a systematic bias in the process of developing and maintaining the ratings at the Moncure and Lillington streamgages, possibly resulting in questionable discharge records?

Discharge measurements made at each of the sites during this period were examined to understand the pattern of percentage differences between the measured discharges and the rating discharges and whether or not a rating shift was applied following the measurements. The trend of percentage differences (whether shifted or unshifted) during the period of analysis also was examined.

Additional information on discharge measurements and ratings analyses is provided in the appendix. Additional information on techniques for collecting and analyzing discharge records can be obtained from Rantz and others (1982, v. 1, 2), Mueller and Wagner (2009), Turnipseed and Sauer (2010), and U.S. Geological Survey (2010b).

Water Use

Water-use information was compiled during the investigation to characterize the number and magnitude of diversions and return flows that occur in the study reach. Because water-use diversions from the main stem have been known to occur, an understanding of the magnitudes of diversions (and the associated consumptive uses) could potentially identify all or part of the suspected flow losses.

After an initial assessment by the North Carolina Division of Water Resources (Don Rayno, North Carolina Division of Water Resources, oral commun., May 28, 2009), five industrial and two municipal facilities with intakes in the Haw or Cape Fear Rivers were identified and contacted in order to obtain detailed information. No major water withdrawals from or return point-source discharges to the Deep River were identified during the study period. Detailed water-use information, particularly with respect to water withdrawals, is commonly limited to hard-copy format when available and requires extensive effort to create an electronic file from paper records or the use of available anecdotal information to compose a general range of daily water use. Both methods were used in this investigation, although the latter method was the more common approach with four of the five industries.

Water-use information also was collected during the series of synoptic measurements to assess the streamflow conditions on each of the six dates. Facility records of water withdrawals (or use) and point-source discharge data were compiled. The study period coincided with a national economic recession, which affected water diversions such that those reported for most of the facilities during the study were lower than in the preceding years.

Of particular note, Shearon Harris Power Plant is located in southwest Wake County (fig. 1) and is adjacent to Shearon Harris Lake, the largest impoundment in the study area. The lake is formed by the impoundment of Buckhorn Creek, the second largest tributary in the study area, and provides cooling water for power generation. The plant currently (2011) consists of one reactor. The 2008–2010 annual water-use reports filed with the North Carolina Division of Water Resources indicated that average daily withdrawal from the lake during these years ranged from 27.9 to 29.1 Mgal/d (43–45 ft^3/s), and the average return point-source discharge to the lake ranged from 15.0 to 17.4 Mgal/d (23–27 ft^3/s; North Carolina Department of Environment and Natural Resources, Division of Water Resources, 2007). Differences between diversions and returns indicate that the average water consumption from Shearon Harris Lake during 2008–2010 ranged from 10.5 to 13.5 Mgal/d (16–21 ft^3/s). Available discharge records from a long-term streamgage on Buckhorn Creek (fig. 1, site 20) downstream from the dam were used to provide insight into the net flow from the drainage basin into the study area. No detailed assessment of the water-use patterns at this facility were deemed necessary because the discharge records available at the streamgage accounted for the contributing streamflow from the basin.

Evaporation

Evaporation from the study reach was estimated during the investigation to determine the portion of suspected flow losses in the study reach that could be attributed to the transformation of water from a liquid state to a gaseous state and removed from the earth's surface. Evaporation is recognized as an important source of water loss during droughts, particularly for operators of large reservoirs managing the balance of inflow to and outflow from impoundments. Losses by evaporation can exceed inflow to impoundments and result in conditions referred to as "negative inflow" to a reservoir. Evaporation from large rivers, such as the Haw and Cape Fear Rivers, likewise can be substantial as water surfaces in wide channels allow for extensive exposure of water to the atmosphere.

The evaporation process occurs in two ways: (1) evaporation of water directly from water bodies, and (2) evapotranspiration (ET) of water through trees and vegetation. Although evaporation occurs throughout the year, the process is cyclical with maximum evaporation occurring during the warmer months. At its peak during the summer months, evaporative losses from the stream channel can be on the order of 0.2–0.3 inch per day (in/d). While this range may not appear substantial at first glance, evaporative losses of this magnitude on a daily basis during the course of a month can well exceed the normal precipitation for an area.

Two methods were used in this study to estimate the portion of flow losses in the study reach that could be attributed to open-water evaporation. The first method involved the compilation and analysis of available pan-evaporation data in the vicinity of the study area. The second method, the Penman-Monteith method (Penman, 1948; Monteith, 1965), was used to compute potential reference-crop ET using meteorological measures for a weather station nearby where climatic data are available. The Penman-Monteith method also can be used to determine reference open-water evaporation with adjustments of coefficients used in the method. It bears noting that evaporation determined by using the Penman-Monteith method is a calculated estimate, not a measured value as with pan-evaporation data.

Evaporation data commonly are collected by using the pan-evaporation method whereby a large stainless steel pan partially filled with a measure of water is placed in an open area and monitored daily to measure changes in water levels, in inches. The placement of the pan in an open area simulates a body of water such as an impoundment. Pan-evaporation rates, however, tend to overestimate actual evaporation from surface water because the pan temperature may be greater than the temperature of a natural water body. Farnsworth and Thompson (1982) determined that evaporation from a shallow lake, wet soil, or other moist natural surface is about 70 percent of the evaporation from a Class A pan under the same meteorological conditions.

Although such data provide a direct measure of evaporative losses, there are only nine National Oceanic and Atmospheric Administration (NOAA) weather stations in North Carolina where pan-evaporation data have been collected. The periods of record vary among the nine stations, and data at these stations generally are intermittent during the periods of record. One NOAA station with data during the investigation period (2008–10) is the Chapel Hill 2W station in Orange County within 30 mi of the study area. The second closest NOAA station with pan-evaporation data during the investigation period is located in Lumberton, Robeson County, more than 60 mi from the study area.

During the study, data from the Chapel Hill 2W NOAA weather station were determined to be the most complete with most daily observations recorded during the warmer months when evaporation is greatest. The Chapel Hill data also were considered to be most representative of evaporation in the study area because of the close proximity of the weather station. Pan-evaporation data for this site date back to March 1949 and were available on an intermittent basis through July 2008 at the time of analysis. The available pan-evaporation data for this station were compiled and analyzed, but because of the intermittent nature of daily pan-evaporation data, monthly median evaporation amounts were determined.

The limited pan-evaporation data near the study area resulted in the determination that use of computed reference-crop ET data based on the Penman-Monteith model could be used to assess evaporative losses from the study reach (Dr. Ryan Boyles, North Carolina State Climatologist, written commun., September 11, 2009). The Penman-Monteith method was specified in a document published by the United Nations Food and Agriculture Organization (FAO; Allen and others, 1998) as the standard method acceptable for estimating reference-crop ET losses for well-watered surfaces (State Climate Office of North Carolina, 2010). The reference surface is defined as a theoretical grass crop. Estimates are computed using meteorological observations of solar radiation, temperature, wind speed, and relative humidity. Using crop coefficients, the determination of ET losses can vary on the basis of the crop of interest and the particular growth stage during the growing season.

The North Carolina State Climate Office has computed and compiled estimated potential daily open-water evaporative losses for NOAA stations across the State using the Penman-Monteith method (State Climate Office of North Carolina, 2010). The daily evaporation amounts were obtained for five NOAA stations in the vicinity of the study area: Chapel Hill-Williams Airport (NOAA station KIGX), Harnett County Airport (KHRJ), Lee County Airport (KTTA), Pope Air Force Base (KPOB) near Fayetteville, and Raleigh-Durham International Airport (KRDU; fig. 8).

The monthly median daily pan-evaporation amounts (adjusted by a pan coefficient of 0.7) for the Chapel Hill 2W station were applied to the estimated surface area of the Haw, Deep, and Cape Fear Rivers between Jordan Lake Dam and Lillington (fig. 1). Daily reference open-water evaporation data compiled for the five nearby NOAA stations using the Penman-Monteith method were averaged and applied to the estimated surface area. The surface area was estimated by creating a trace of river boundaries using computerized topographical maps. Applying the daily evaporation amounts to the surface area resulted in an estimated volume of water removed daily from the rivers.

Groundwater and Surface-Water Interaction

From July 2009 to November 2010, several field investigative and monitoring methods were applied to describe the groundwater and surface-water interaction within the flood plain of the Cape Fear River. Elements of the study included a geophysical survey, a distributed temperature sensing survey and an aerial infrared survey to help define the groundwater and surface-water interaction along the study reach. Continuous groundwater monitoring also was conducted at two piezometer transects on the Cape Fear River.

Electromagnetic Geophysical Survey

A GEM 2 digital, multifrequency electromagnetic (EM) sensor was used during July and August 2009 to collect a continuous EM profile along the Haw, Deep, and Cape Fear Rivers within the study area. The focus of the survey was to identify locations of possible diabase dikes (fig. 5) intersecting the bottom of the river.

The GEM 2 sensor has an electric coil that transmits a waveform at frequencies ranging between 300 and 24,000 hertz (Hz) that induces a magnetic field in susceptible rocks, such as diabase. The sensor receiver coil then measures the electrical conductivity of the response. Because of the multifrequency capability of the GEM 2, the sensor has the ability to penetrate to depths beneath the surface depending on the rock type encountered. At low frequencies, the waveform produced by the GEM 2 penetrates deeper into the earth, and the response is comparable to the response of a magnetometer (Geophex, Ltd., 2011).

The GEM 2 sensor was equipped with a global positioning system (GPS) tracking device and towed behind a boat for approximately 53 mi along the main stem and the Deep River to trace variations or anomalies in the magnetic signal. Two 26.5-mi-long survey lines were completed by towing the instrument along each side of most of the channel. Following the survey, the USGS Office of Groundwater Branch of Geophysics provided processing support to filter the signal data (Troy R. Brosten, U.S. Geological Survey, written commun., 2009). Raw data logged by the GEM 2 were recorded in sensor-specific units of parts per million. While multiple frequencies were collected during the survey, the frequency that displayed the deepest resolution for each reach was selected for data analysis. In the reach above Buckhorn Dam, the 450-Hz frequency was selected, and a frequency of 1,170 Hz was selected for the lower reach. A lower frequency was needed in the reach above Buckhorn Dam in order to penetrate the thick water column (greater than 20 ft in places) in the Cape Fear River.

Groundwater Temperature

Additional characterization of the interaction between groundwater and surface water was conducted by evaluating thermal properties within the hyporheic zone of the Cape Fear

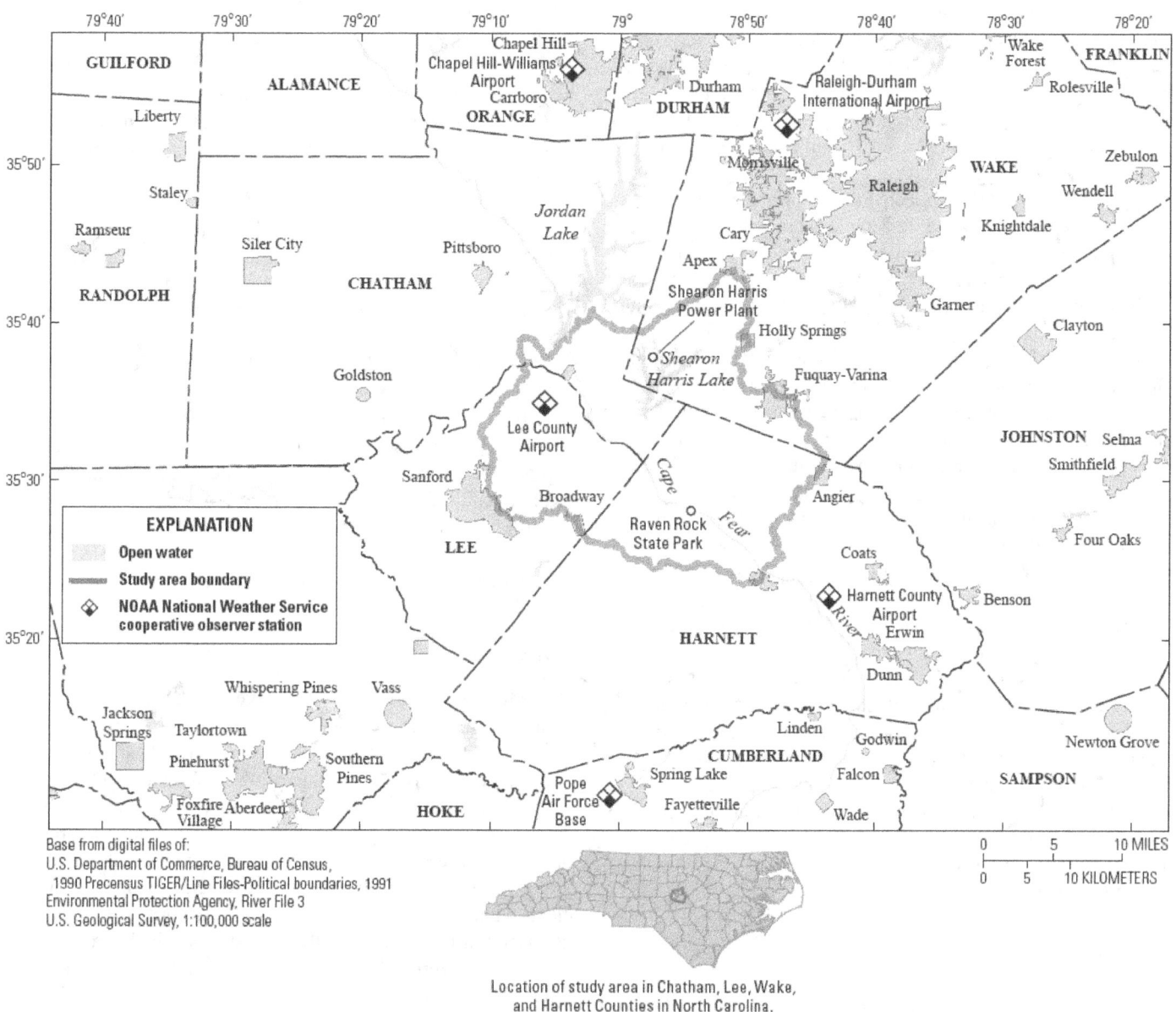

Figure 8. Locations of nearby National Oceanic and Atmospheric Administration National Weather Service cooperative observer stations where daily reference, open-water evaporation data were compiled for the study based on the Penman-Monteith method.

River. Two novel approaches, distributed-temperature sensing and thermal infrared imaging, were applied during this study to characterize the spatial and temporal variation of groundwater discharge.

Distributed-Temperature Sensing (DTS) Survey

Fiber-optic distributed-temperature sensing (FO-DTS) is a relatively new technology that can measure temperature with a spatial resolution of less than 1 yard along a fiber-optic cable that may be a mile or more long. Temperature precision of 0.1 degree Celsius (°C) and a temporal resolution of 90 seconds are attainable using FO-DTS technology. Physical principles of FO-DTS technology are described in Selker and others (2006). In this study, temperature data were collected using a SensorTran FO-DTS.

A 3,000-ft-long shielded telecommunication fiber-optic cable was deployed in a reach of the Cape Fear River near Raven Rock State Park during August 12–14, 2009 (fig. 1, site DTS). The fiber-optic cable was laid temporarily on the bed of the Cape Fear River in two lines parallel to the river-bank to measure differences in temperature along the lines over a 36-hour period, August 11–13, 2009. Temperature data were collected every 2.75 minutes at intervals of about 1.7 ft along the length of the fiber-optic cable.

Thermal Infrared (TIR) Survey

Thermal infrared (TIR) imaging has been used to remotely sense groundwater discharge and provide a non-invasive screening tool for the identification of groundwater seeps over a large geographic area (Banks and others, 1996). To locate areas of seepage that contribute groundwater to the Cape Fear River, a high-resolution, low-altitude airborne TIR survey was performed during the early-morning hours on February 27, 2010. The 24.5-mi reach of the Haw and Cape Fear Rivers extending from B. Everett Jordan Dam to Lillington was investigated using TIR.

A USGS contractor collected the TIR imagery using a FLIR® Systems, Inc. digital infrared camera with 1 megapixel resolution (1024 x 1024) mounted on a fixed wing aircraft. The camera measured radiant energy (in the infrared wave band) emitted from the water surface and stored the readings as values of pixels in a digital image, whereas a standard thermometer measures kinetic energy through direct contact with an object or fluid. In a TIR image, land and water features are distinguishable because of differences in radiant temperature. To minimize interference caused by reflected solar thermal energy, the TIR survey was conducted in the pre-dawn hours during a winter month when the temperature of the groundwater was warmer than the ground surface and surface water. Additionally, the air temperature did not exceed 3 °C and no measurable precipitation occurred within the preceding 36-hour period at the KRDU weather station (Raleigh-Durham International Airport, fig. 8). The river surface was not frozen, and the instantaneous gage height measured at the Cape Fear River at Lillington (fig. 1, site 28) did not exceed 4.0 ft at the time of image collection.

The TIR imagery was collected in one flight using a series of passes to gather enough overlapping imagery to cover the entire river surface plus a 50-ft buffer on each side of the river. A georeferenced photo mosaic of overlapping digital TIR imagery was assembled by the contractor at the end of imagery collection. Groundwater seepage was depicted on the images as lighter (warmer) areas compared to darker (cooler) surrounding areas.

Piezometer Transects

Piezometer transects were installed at two locations on the Cape Fear River to monitor water levels in the flood plain and river (fig. 1, PZ1, PZ2) and determine the variability of groundwater and surface-water interaction. A piezometer is a well that has a short screen so the water level represents hydraulic head in only a small part of the groundwater system (Winter and others, 1998). Clustering a group of piezometers at different depths at the same location is referred to as a piezometer nest (or transect) and allows data collection that can be used to interpret the movement of water within the transect.

To install the piezometers, individual boreholes were excavated in the flood-plain and river-bed sediments by jetting, using a high-velocity stream of water either to a specific depth or to resistance by consolidated material. A centrifugal pump supplied a pressurized stream of water through a jetting tube to excavate a 6-inch (in.) diameter hole and remove the liquefied sediment. The Cape Fear River was the water source used for jetting. Because of the cohesive nature of the floodplain sediments, a temporary casing was not needed to prevent sediments from collapsing back into the hole.

After each borehole was excavated to the prescribed depth, a 1.5-in. diameter schedule 40 polyvinyl chloride (PVC) piezometer casing was placed in the borehole. All of the piezometers were constructed with a 0.01-in. machine-slotted well screen with a 1-ft length. Clean medium-sized (no. 2) filter sand was poured into the annular space between the PVC casing and the borehole wall to a level approximately 0.25 ft above the top of the well screen. A seal of bentonite pellets, hydrated with water obtained from the Cape Fear River, was placed on top of the sand filter pack to the floodplain or river-bed surface. After installation, the altitude of each piezometer was determined by using traditional survey techniques referenced to the altitude of a nearby temporary benchmark established by using GPS satellites. Construction details for each piezometer are listed in table 2.

In order to monitor horizontal and vertical groundwater-flow paths in the flood plain, the piezometers were installed along a high- to low-topographic profile perpendicular to the river edge and following a presumed flow path from recharge (flood-plain terrace) to discharge areas (river; fig. 9). In the riverbed, a nest of piezometers was installed at increasing depths to monitor water levels (vertical gradients).

Continuous Monitoring

Monitoring of groundwater level and river stage was conducted at each piezometer transect. Groundwater-level and river-stage data were collected on a continual basis by using an absolute (nonvented) self-logging submersible pressure transducer. Consequently, groundwater-level and river-stage data were corrected using barometric pressure data recorded onsite. The pressure transducers were field checked every 4 to 6 weeks and corrected to measurements made with an electric tape to ensure accurate reading, according to methods described in Freeman and others (2004). Vertical and horizontal gradients were calculated by comparing water-level altitudes. Continuous and periodic groundwater-level and river-stage data are stored in the USGS National Water Information System (NWIS) database (U.S. Geological Survey, 2001).

Results and Interpretation of Data

The multidiscipline data collection and analyses completed for this investigation revealed that flow losses occur in the reach of the Cape Fear River between Jordan Lake Dam and the USGS streamgage at Lillington (fig. 1, site 28). The results of this investigation are discussed in the following sections.

Surface Water

Several methods were applied during the study to aid in understanding the occurrence of suspected flow losses and to identify possible causes. Examination of the streamflow records collected along the study reach coupled with six series

Table 2. Characteristics of the piezometers and adjacent surface-water sites on the Cape Fear River in Chatham and Harnett Counties, North Carolina.

[NAVD 88, North American Vertical Datum of 1988; in., inch; PZ, piezometer; °, degree; ', minute; ", second; PVC, polyvinyl chloride, schedule 40 casing; H, hyporheic; R, river; FP, flood plain; NA, not applicable]

Site identification number	Station name	Latitude	Longitude	Construction date	Land-surface elevation (feet above NAVD 88)	Top of casing elevation (feet above NAVD 88)	Casing material	Casing diameter (inches)	Screened interval (feet below land surface) From	To	Screen type	Zone monitored
PZ1, Regional power plant transect (profiles shown in fig. 9A)												
353304079013401	CH-231 Cape Fear PE-1U	35°33'04.1"	79°01'34.5"	Oct. 1, 2009	156.71	162.75	PVC	1.5	0.96	1.96	0.01-in. slotted PVC	H
353304079013402	CH-232 Cape Fear PE-1M	35°33'04.1"	79°01'34.5"	Oct. 1, 2009	156.73	162.47	PVC	1.5	3.98	4.98	0.01-in. slotted PVC	H
353304079013403	CH-233 Cape Fear PE-1D	35°33'04.1"	79°01'34.5"	Oct. 1, 2009	156.69	162.61	PVC	1.5	6.98	7.98	0.01-in. slotted PVC	H
353304079013404	CH-234 Cape Fear PE-1R	35°33'04.1"	79°01'34.5"	Oct. 1, 2009	156.65	162.57	PVC	1.5		0.00	0.01-in. slotted PVC	R
353304079013405	CH-235 Cape Fear PE-2	35°33'04.1"	79°01'34.5"	Oct. 1, 2009	159.26	163.15	PVC	1.5	5.84	6.84	0.01-in. slotted PVC	FP
353304079013406	CH-236 Cape Fear PE-3	35°33'04.1"	79°01'34.5"	Oct. 1, 2009	159.98	163.65	PVC	1.5	4.59	5.59	0.01-in. slotted PVC	FP
PZ2, Bradley Road transect (profiles shown in fig. 9B)												
352630078511401	HR-060 Cape Fear BR-1D	35°26'29.4"	78°51'14.9"	Aug. 11, 2009	107.01	113.96	PVC	1.5	2.75	3.75	0.01-in. slotted PVC	H
352630078511402	HR-061 Cape Fear BR-1U	35°26'29.4"	78°51'14.9"	Aug. 11, 2009	106.80	113.80	PVC	1.5	0.50	1.50	0.01-in. slotted PVC	H
352630078511403	HR-062 Cape Fear BR-1R	35°26'29.4"	78°51'14.9"	Aug. 11, 2009	106.37	113.8	PVC	1.5		0.00	0.01-in. slotted PVC	R
352630078511404	HR-063 Cape Fear BR-2	35°26'29.5"	78°51'14.9"	Aug. 11, 2009	107.76	114.21	PVC	1.5	3.25	4.25	0.01-in. slotted PVC	H
352630078511405	HR-064 Cape Fear BR-3	35°26'29.5"	78°51'14.8"	Aug. 11, 2009	112.57	115.17	PVC	1.5	7.15	8.15	0.01-in. slotted PVC	FP
352630078511406	HR-065 Cape Fear BR-4	35°26'29.6"	78°51'14.8"	Aug. 11, 2009	114.23	117.91	PVC	1.5	6.07	7.07	0.01-in. slotted PVC	FP
0210215960 (site 14, fig. 6)	Cape Fear River above NC Highway 42 near Corinth, NC	35°33'02.68"	79°01'36.42"	Aug. 11, 2009	156.70	NA	NA	NA	NA	NA	NA	R
02102283 (site 25, fig. 7)	Cape Fear River adjacent to Bradley Road near Lillington, NC	35°26'28.92"	78°51'16.85"	Apr. 27, 2009	106.75	NA	NA	NA	NA	NA	NA	R

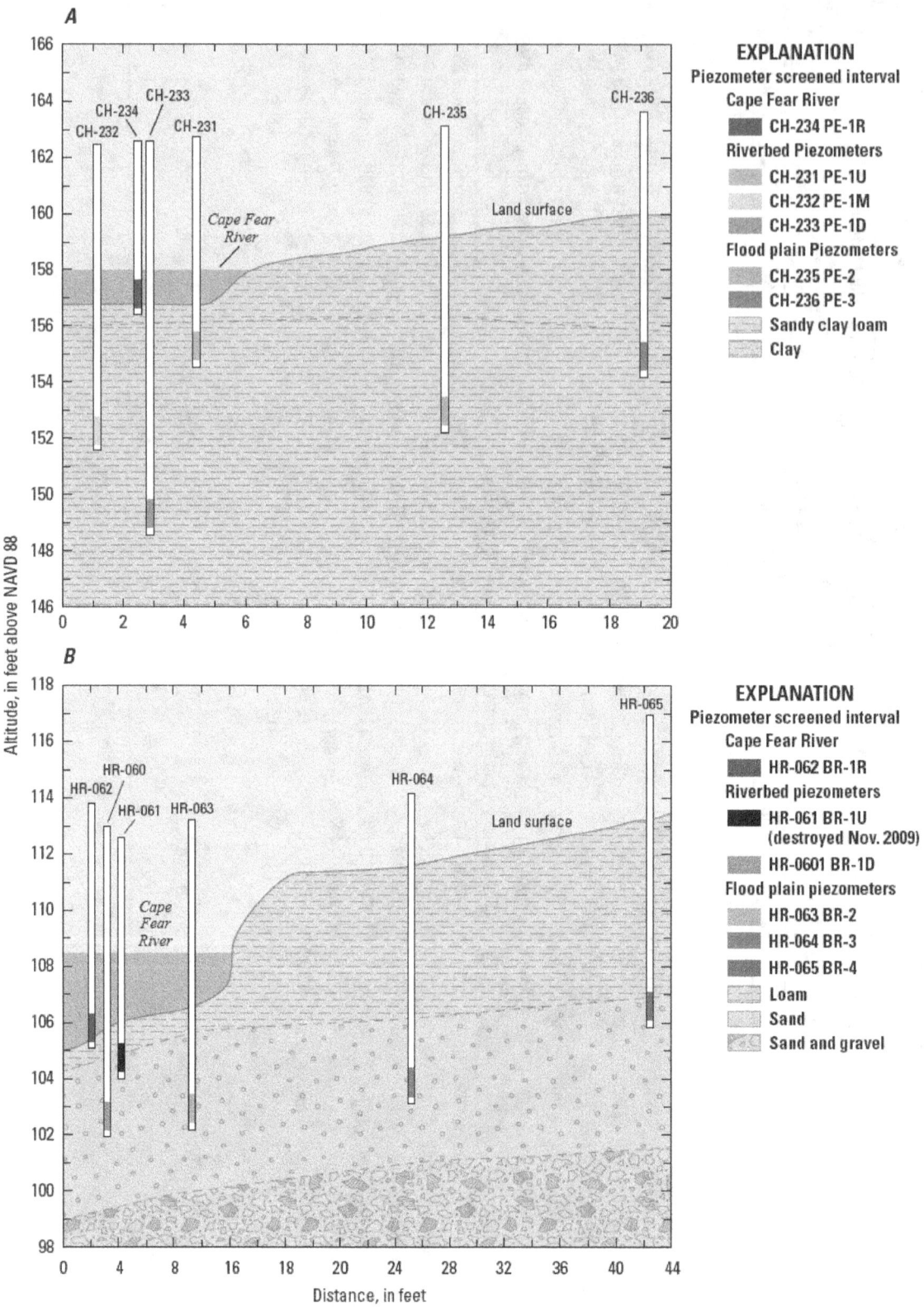

Figure 9.   Piezometer profiles and soils encountered at the (*A*) regional power plant transect (PZ1) near Corinth and (*B*) Bradley Road transect (PZ2) near Lillington, North Carolina (sites shown in fig. 1).

of synoptic discharge measurements provided information concerning the magnitudes of flow losses and whether or not the losses consistently occurred at a given location. Water use was examined and open-water evaporative losses were estimated, both of which were determined to cause flow losses in the study area. Review of the measured discharges and ratings at the two long-term streamgages at Moncure and Lillington did not indicate any systematic bias in the stage-discharge ratings used to compute discharge records. Results of the surface-water analyses are provided in the following sections.

Flow Loss Analyses

The first analyses in the study focused on developing an understanding of the magnitude and frequency of flow losses by using flow-release data reported by the USACE from Jordan Lake Dam and discharge records at three USGS streamgaging stations (fig. 1, sites 3, 20, 28). Recent and historical records based on active and discontinued stations were examined to document the differences in the number of days of flow loss in varying periods of record. The flow loss analyses of daily discharge records confirmed the flow loss between the upstream and downstream ends of the study reach but did not provide insight into possible causes of the losses.

For the purposes of this study, the quantification of flow losses was determined by summing the daily discharges at Jordan Lake Dam, Deep River at Moncure, and Buckhorn Creek near Corinth (fig. 1, sites D1, 3, 20, respectively) and subtracting this value from the discharge at Cape Fear River at Lillington (site 28). As defined in this report, positive net differences indicate a gain in flow between the upstream inputs and the streamgage at Lillington, whereas negative net differences indicate a possible flow loss between these locations. Net differences in flow were determined for a 1-day time scale using daily mean discharge or flow release and for a 7-day time scale using 7-day average discharge or flow release.

Three flow comparisons were completed to evaluate periods of potential flow loss using records for the 1983–2010 water years as well as two comparisons based on records from January 1963 to December 1970 (prior to Jordan Lake construction) for discontinued streamgages (table 3). The period 1963–70 was specifically chosen because it corresponds to the time after active regulation by Buckhorn Dam ceased but prior to the start of construction on Jordan Lake.

The first historical flow comparison was made using data from the discontinued streamgage on the Haw River near Pittsboro (USGS station 02097000; drainage area 1,310 mi²), which was operated October 1928–September 1973 and discontinued during the construction of Jordan Lake because the streamgage was located in the inundation area. The second historical flow comparison was made using data from the discontinued streamgage on the Haw River near Haywood (USGS station 02098200; drainage area approximately 1,700 mi²), which was operated October 1965–September 1978. The location of the discontinued streamgage near Haywood is downstream from the current streamgage

below Jordan Lake Dam (fig. 1, site 1). In each historical comparison, the discharges at the discontinued streamgages were summed with the discharges at Deep River at Moncure (site 3) and compared with corresponding discharges at Cape Fear River at Lillington (site 28). Among the 2,922 possible days for comparison during January 1963–December 1970, a total of 2,922 and 1,918 days, respectively, were available for flow comparisons 1 and 2 for all three streamgages. The smaller number of days available in flow comparison 2 is because the period of record for the Haywood streamgage began in October 1965.

Flow comparisons 3, 4, and 5 were completed using data for the 1983–2010 water years. This period includes the presence of Jordan Lake and Shearon Harris Lake, which affect part of the input flows into the study area. Input flows for flow comparisons 3 and 5 were daily flow releases reported for Jordan Lake Dam (fig. 1, site D1) and the discharges at Deep River at Moncure (site 3) and Buckhorn Creek near Corinth (site 20). The differences between flow comparisons 3 and 5 were that input flows were compared to same-day and next-day (1-day lag) discharges, respectively, at the Lillington streamgage (site 28). Completing a comparison based on next-day discharges at the Lillington streamgage allowed for consideration of travel time in flow between Jordan Lake Dam and Lillington. Flow comparison 4 was made using data from the USGS streamgage on the Haw River below Jordan Lake (site 1) for one of the input flows in place of flow releases reported for the dam; however, comparisons based on data from the USGS streamgage below Jordan Lake are limited in terms of available record (1983–92 water years). As discussed in subsequent sections, comparisons using data collected during the 2008–2010 water years at this streamgage could not be completed because of technical concerns with the velocity data collected during the study. Nevertheless, the occurrence of flow loss noted in this comparison reinforces the recognition that flow loss is part of the overall flow dynamics in this reach.

For each flow comparison, the number of days of flow loss were tabulated (table 3) for two categories: the number of days of flow loss during the period of interest regardless of the preceding flow conditions upstream from the study area and the number of days of flow loss based on a filtered comparison in which daily records for flow releases at Jordan Lake Dam (site D1) and discharges at Deep River at Moncure (site 3) both changed less than 5 percent from the previous day's records. The filter was applied to examine the days when little to no temporal change in flow occurred, indicating that no peak flow was passing through the study area along the main stem. For the number of days of flow loss filtered, the minimum, median, and maximum flow losses were tabulated (table 3).

In all comparisons, the number of days of flow loss based on a filtered assessment is substantially less than the number of days of flow loss based on a nonfiltered assessment (table 3). For example, in flow comparison 3, among 10,227 days during the 1983–2010 water years, flow loss occurred on a total of 2,941 days (28.8 percent) based on

Table 3.   Number of days with flow loss and flow-loss statistics for the Haw, Deep, and Cape Fear Rivers between Jordan Lake Dam and Lillington, North Carolina, for the periods January 1963 to December 1970 and the 1983 - 2010 water years.

[USGS, U.S. Geological Survey; ft³/s, cubic foot per second; flow comparisons discussed in detail in report. Number of days flow loss filtered refers to days in which changes in flow at upstream sites were less than 5 percent from previous day.]

Flow loss based on daily mean discharge					Flow loss based on 7-day average discharge				
Number of days flow loss and percentage of days with flow at all sites	Number of days flow loss filtered and percentage of days with flow at all sites	Minimum, in ft³/s (date)	Median, in ft³/s	Maximum, in ft³/s (date)	Number of days flow loss and percentage of days with flow at all sites	Number of days flow loss filtered and percentage of days with flow at all sites	Minimum, in ft³/s (date)	Median, in ft³/s	Maximum, in ft³/s (date)
Flow comparison 1: Haw River near Pittsboro (USGS station 02097000) plus Deep River at Moncure (site 3) compared to Cape Fear River at Lillington (site 28), 2,922 days with flow at all 3 sites among 2,922 possible days during period of interest (January 1963 through December 1970)									
391	3	1.0	5.0	26.0	179	33	0.14	5.1	78.1
13.4%	0.1%	(4/24/1968)		(10/11/1968)	6.1%	1.1%	(9/18/1968)		(8/17/1966)
Flow comparison 2: Haw River near Haywood (USGS station 02098200) plus Deep River at Moncure (site 3) compared to Cape Fear River at Lillington (site 28), 1,918 days with flow at all 3 sites among 2,922 possible days during period of interest (January 1963 through December 1970)									
417	17	1.0	33.0	100	318	90	1.1	16.4	740
21.7%	0.9%	(9/21/1968)		(2/1/1966)	16.6%	4.7%	(9/29/1968)		(3/19/1968)
Flow comparison 3: Jordan Lake Dam flow releases (site D1), Deep River at Moncure (site 3), and Buckhorn Creek near Corinth (site 20) compared to Cape Fear River at Lillington (site 28), 10,227 days with flow at all 4 sites among 10,227 possible days during period of interest (1983-2010 water years)									
2,941	408	0.49	37.2	2,150	2,350	647	0.07	37.4	1,450
28.8%	4.0%	(9/11/1989)		(4/24/1998)	23.0%	6.3%	(11/28/2007)		(4/25/2003)
Flow comparison 4: Flows at Haw River below Jordan Lake dam (site 1), Deep River at Moncure (site 3), and Buckhorn Creek near Corinth (site 20) compared to Cape Fear River at Lillington (site 28), 2,069 days with flow at all 4 sites among 3,653 possible days during period of interest (1983-1992 water years)									
1,182	169	1.8	92.0	433	1,111	296	0.52	85.2	1,030
57.1%	8.2%	(11/23/1984)		(10/24/1988)	53.7%	14.3%	(11/27/1984)		(3/7/1989)
Flow comparison 5: Jordan Lake dam flow releases (site D1), Deep River at Moncure (site 3), and Buckhorn Creek near Corinth (site 20) compared to next-day Cape Fear River at Lillington (site 28) in effort to look at possible losses on a 1-day adjusted time scale, 10,226 days with flow at all 4 sites among 10,227 possible days during period of interest (1983-2010 water years)									
3,775	394	0.18	43.2	3,330	3,191	656	0.09	38.0	1,900
36.9%	3.9%	(8/26/2008)		(11/25/2006)	31.2%	6.4%	(7/11/2009)		(4/26/2003)

comparison of the daily mean discharge between the inputs and the Lillington streamgage (site 28). However, flow loss occurred on 408 days (4.0 percent) when conditions were relatively steady with respect to the previous day's records. The flow loss during these 408 days ranged from 0.49 to 2,150 ft³/s, with a median flow loss of 37.2 ft³/s (table 3). In terms of 7-day average discharge, the number of days of flow loss for this comparison was determined to be 2,350 days (23.0 percent) and 647 days (6.3 percent) for the nonfiltered and filtered assessments, respectively. The flow loss on these 647 days ranged from 0.07 to 1,450 ft³/s, with a median flow loss of 37.4 ft³/s (table 3). An example of flow loss can be seen in the daily and unit hydrographs for October 5–14, 2007, a period during which drought conditions occurred (fig. 10). Both daily and 7-day flow losses occurred on all days during this period, and daily flow losses were computed.

Examination of the maximum flow loss amounts, particularly for flow comparisons 3 and 5 (table 3), indicates the filtered assessments did not filter out all nonpeak-flow conditions, because flow losses at this magnitude are not true flow losses. The maximum flow loss determined for comparison 3 highlighted an interesting occurrence in the flow patterns at the Lillington streamgage where flows decreased and then increased again by about 5,000 ft³/s within a 36-hour period during April 23–25, 1998 (fig. 11). Limited information is available to fully explain this unusual flow pattern, suspicion of possible sudden changes in flow release were confirmed by the USACE (Terry Brown, U.S. Army Corps of Engineers (retired), written commun., December 16, 2009). Of note, the lake elevation for Jordan Lake on April 21, 1998, exceeded 220 ft, more than 4 ft above the guide curve elevation of 216 ft used in daily lake-management operations (fig. 11A).

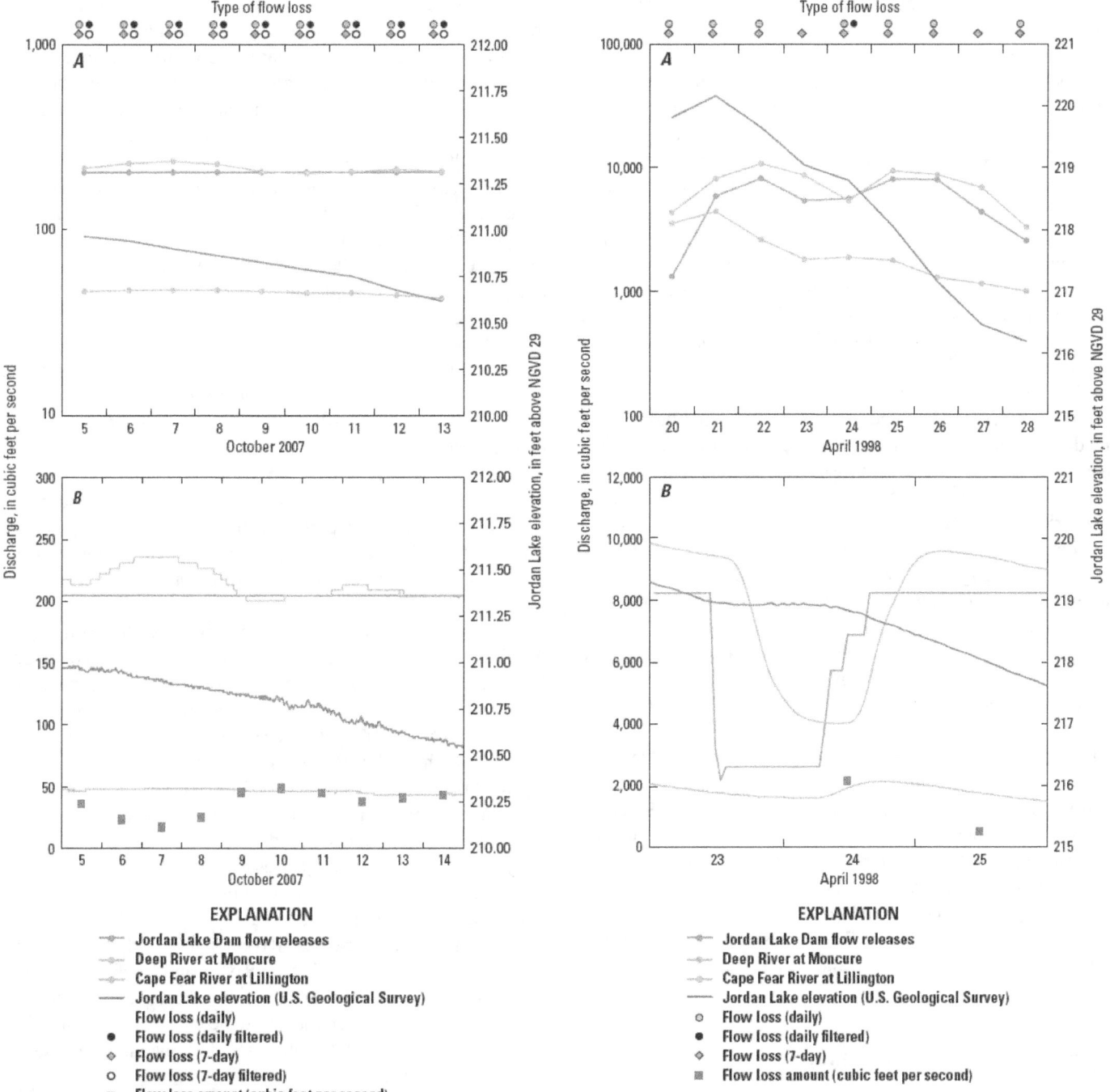

Figure 10. (*A*) Daily mean and (*B*) instantaneous discharges in the Cape Fear River and tributaries, North Carolina, during October 5–14, 2007, with daily and 7-day flow losses based on daily discharge flow comparison 3.

Figure 11. (*A*) Daily mean and (*B*) instantaneous discharges for days prior to and following April 24, 1998, on which maximum flow loss occurred in the Cape Fear River and tributaries, North Carolina, based on daily discharge flow comparison 3.

An elevation at this level typically results in actions to reduce the lake level as rapidly and safely as possible, which is reflected in the daily hydrograph for April 21–22, 1998. The lake elevation stabilized between the mid-day hours on April 23 and 24, suggesting a sudden and substantial change in flow releases from Jordan Lake Dam (fig. 11*B*). No hourly flow data were initially available for this period to confirm the sudden change in discharge at Lillington. Flow releases estimated by the USACE (Ashley Hatchell, U.S. Army Corps of Engineers, written commun., June 28, 2011), however, confirm the sudden and substantial changes in the flow releases on April 23–24. Information provided with the estimated flow releases also included a historical note referencing heavy rain in the lower part of the Cape Fear River basin, indicating that the sudden changes may have been made to help mitigate high flows in the lower part of the basin.

Flow losses determined in flow comparison 3 based on the daily mean discharges were selected for further inspection using statistical histograms to assess the frequency of several factors related to the flow loss determined in the filtered assessment (fig. 12). Approximately 61 percent of the flow loss days occurred when the daily mean discharge at the Lillington streamgage (site 28) was at the target flow of 600 ft^3/s or less. In terms of flow loss amounts, the histogram indicates that approximately 64 percent of the flow losses are 50 ft^3/s or less. Expressed as a percentage of the daily mean discharge at the Lillington streamgage, nearly 68 percent of the flow losses constitute 10 percent or less of the Lillington flows. In terms of days of occurrence during the calendar year, about 82 percent of the flow loss days were during the months of May through October. The months with the highest number of days of flow loss were June (16.7 percent), September (16.9 percent), and October (19.4 percent). Information gleaned from the histograms indicates the majority of flow losses occurred during the warm season months, particularly during the base-flow months of September and October. The histograms also indicate that a majority of the flow losses accounted for less than 10 percent of the daily mean discharge at the Lillington streamgage.

Synoptic Discharge Measurements

A series of synoptic discharge measurements were made at selected locations along the main stem between Jordan Lake Dam (fig. 1, site D1) and the Lillington streamgage (site 28) on six different occasions in 2009 (fig. 13). The purpose of each series of synoptic measurements was to create a "snapshot" of the flow conditions along the main stem to determine if a particular pattern could be detected that might identify a consistent flow-loss location, thereby leading to further investigation into the cause(s) for flow loss at that location. Because the flow analyses indicated the majority of flow losses occurred under base-flow conditions during the warmer months, all synoptic-measurement series except the April 27 series were conducted under these conditions on June 25, July 22, August 18, September 9, and October 1, 2009

(figs. 14–19; tables 4–9). The completion of each synoptic-measurement series indicated that the largest change in the main-stem flow occurs immediately downstream from the regional power-plant intake (fig. 1, site F7). However, the synoptic measurements did not indicate any other location where another previously unknown flow loss was observed consistently on each date.

Discharge measurements were made by using acoustic measuring equipment and USGS techniques (Mueller and Wagner, 2009; U.S. Geological Survey, 2010b). For each measured discharge collected during the synoptic series on the main stem and the Deep River, the velocity, standard deviation, 95-percent confidence interval, and percentage of change in discharge from the next upstream discharge on the main stem are summarized in tables 4–9. The standard deviation and measured discharge were used to compute the coefficient of variation, which was used with the number of passes completed during the measurement across the section (typically four or eight) to compute a 95-percent confidence interval of measurement uncertainty. Following the review procedures, the discharge measurements for three sites (6, 7, and 14) on June 25 (table 5) and one site (17) on October 1 (table 9) were deemed unreliable and, therefore, were not used in further analyses.

The first series of synoptic measurements made on April 27 were not at base-flow conditions but provided insight in preparing for the remaining five series of synoptic measurements. The discharges at the Lillington streamgage ranged from 300 to 600 ft^3/s for the five series of measurements made during June through October and was about 1,400 ft^3/s for the April 27 measurements (fig. 13). Known diversions for water use—both withdrawals and return point-source discharges—were compiled and documented when available for each series of synoptic measurements (tables 4–9).

The measured discharges highlight some challenges associated with attempting to create a profile of discharges along the Haw and Cape Fear Rivers, particularly between the confluence of the Haw and Deep Rivers and Buckhorn Dam. Under ideal circumstances, the measured discharges would be expected to accurately characterize the changes in discharges due to confluences with tributaries and to diversions along a given riverine reach.

Among the series of six synoptic measurements, the sum of the measured discharges at site 2 on the Haw River and site 4 on the Deep River (tables 4–9) were +5.5, +3.7, –3.7, –20, –0.3, and –13.9 percent different from the measured discharge at site 5 on the Cape Fear River just downstream from the confluence and upstream from the intake at site F7 (fig. 6). Assessment of the 95-percent confidence intervals for the measured discharges at these three sites indicated that the interval of measurement uncertainty was greater than 8 percent for part of or all three sites on each of the synoptic-measurement dates with the exception of April 27, which was the date with the highest range of flows measured (fig. 13). As indicated in the surface-water quality-assurance plan for the USGS North Carolina Water Science Center (U.S. Geological Survey, 2010b),

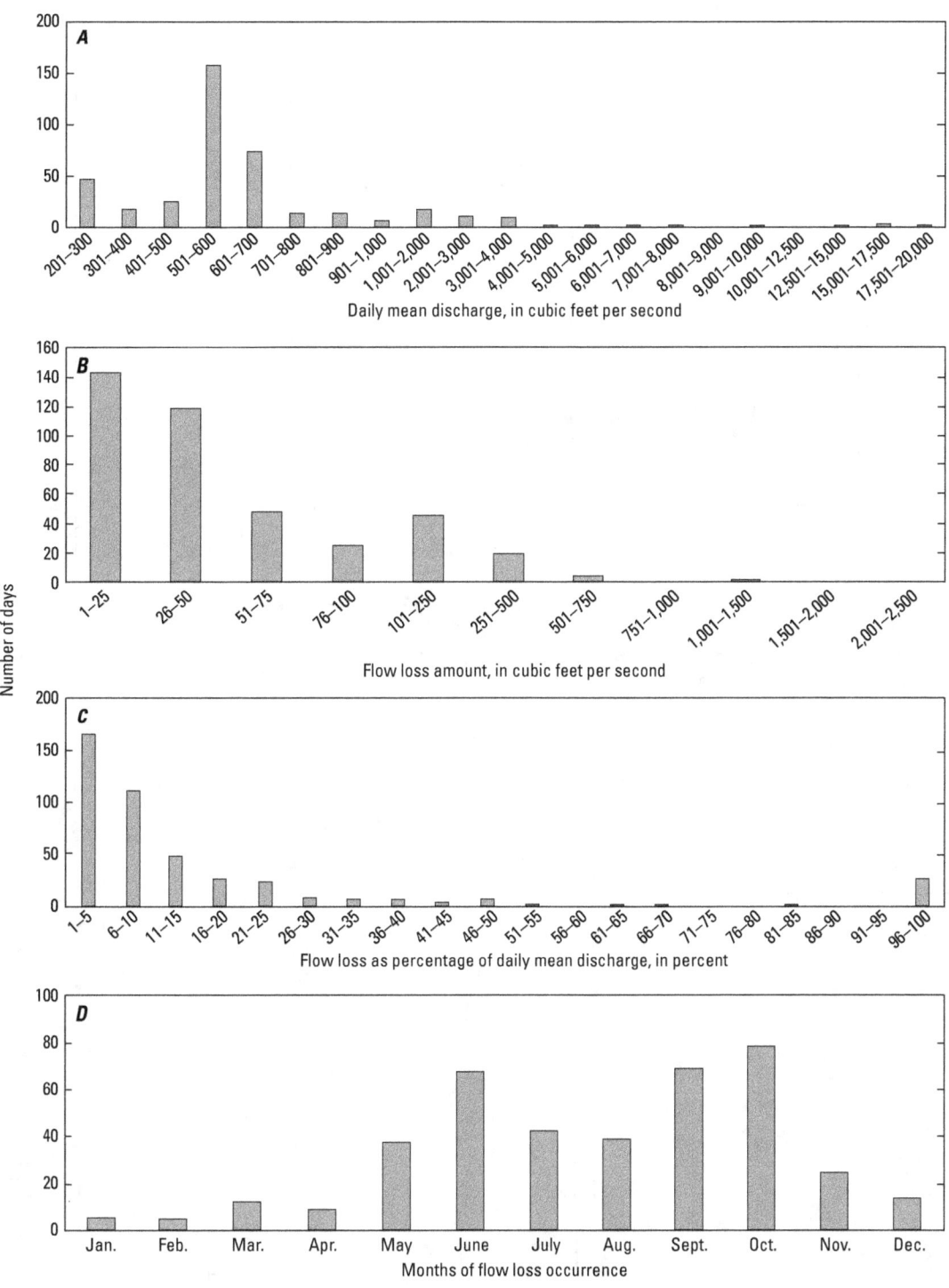

Figure 12.   Filtered flow-loss occurrences in the Cape Fear River determined in flow comparison 3 based on (*A*) daily mean discharge, (*B*) flow-loss amount, (*C*) percentage of daily mean discharge, and (*D*) month of occurrence.

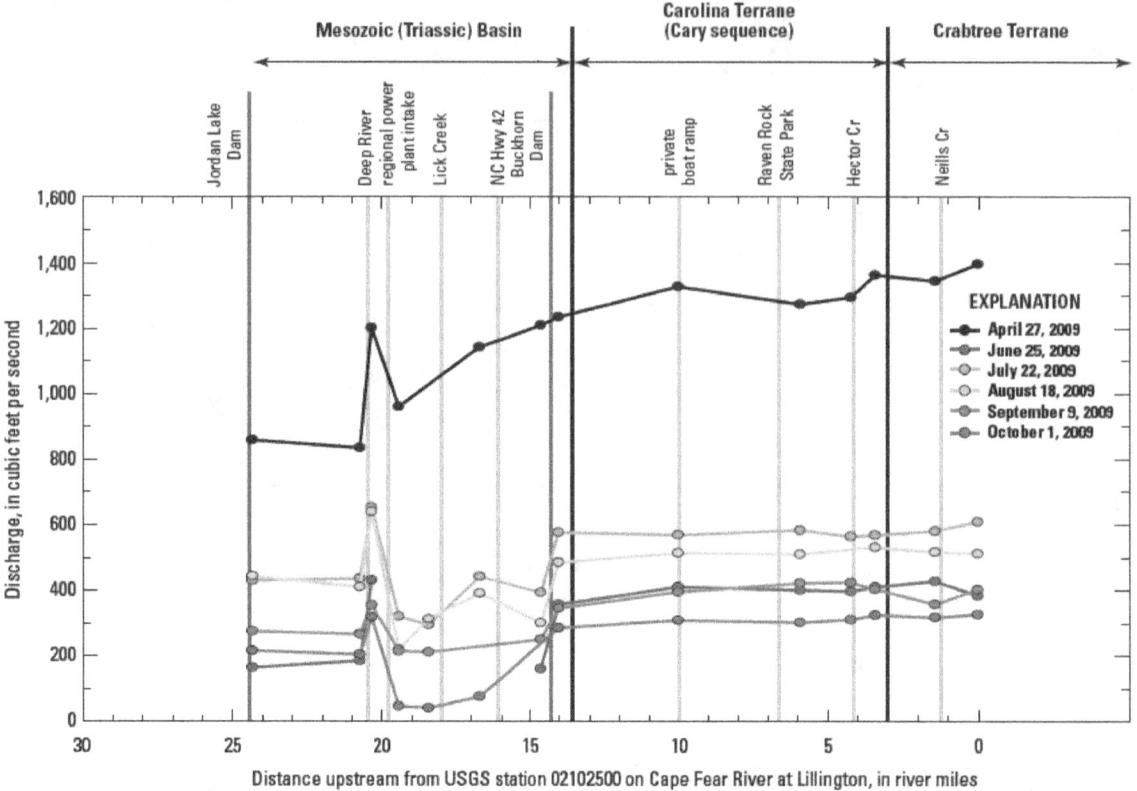

Figure 13. Discharges measured during a series of six synoptic discharge measurements in 2009 on the Haw and Cape Fear Rivers between Jordan Lake Dam and Lillington, North Carolina.

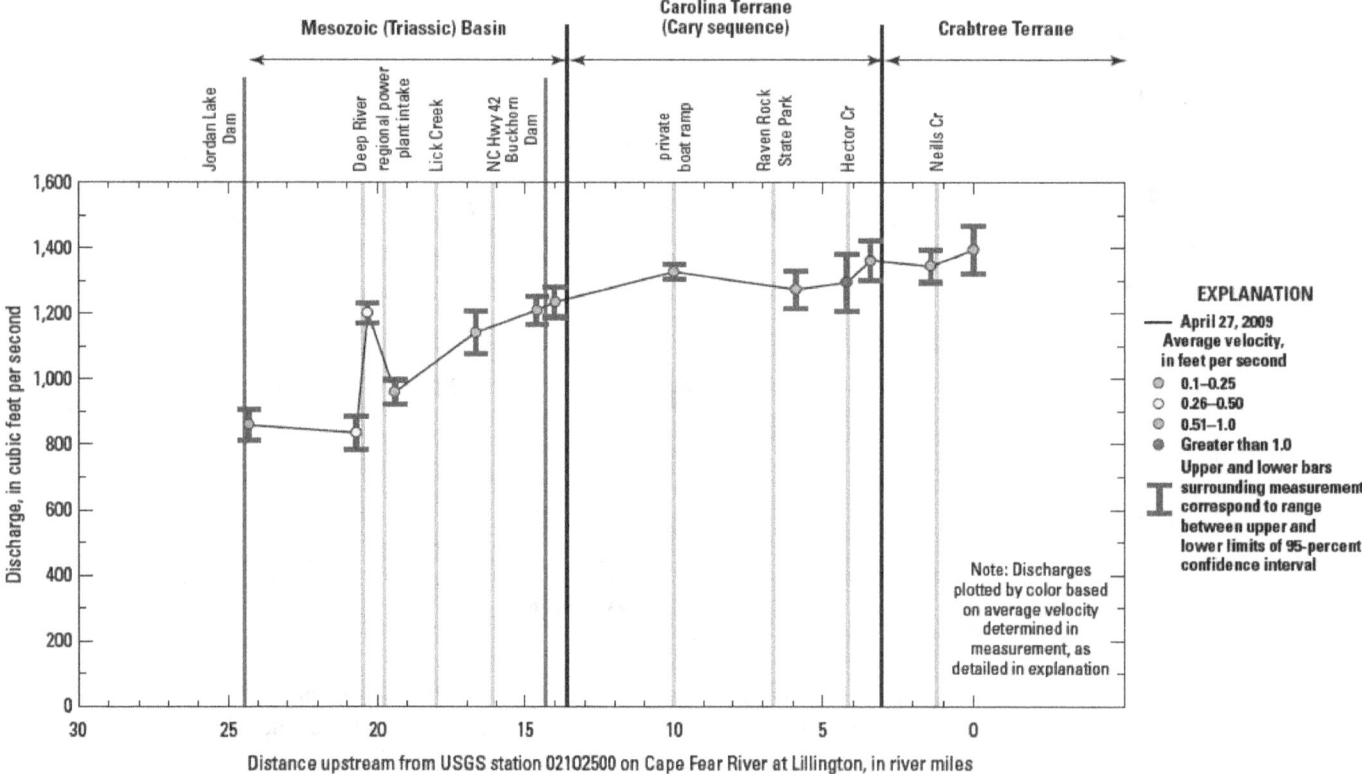

Figure 14. Measured discharges on April 27, 2009, on the Haw and Cape Fear Rivers between Jordan Lake Dam and Lillington, North Carolina.

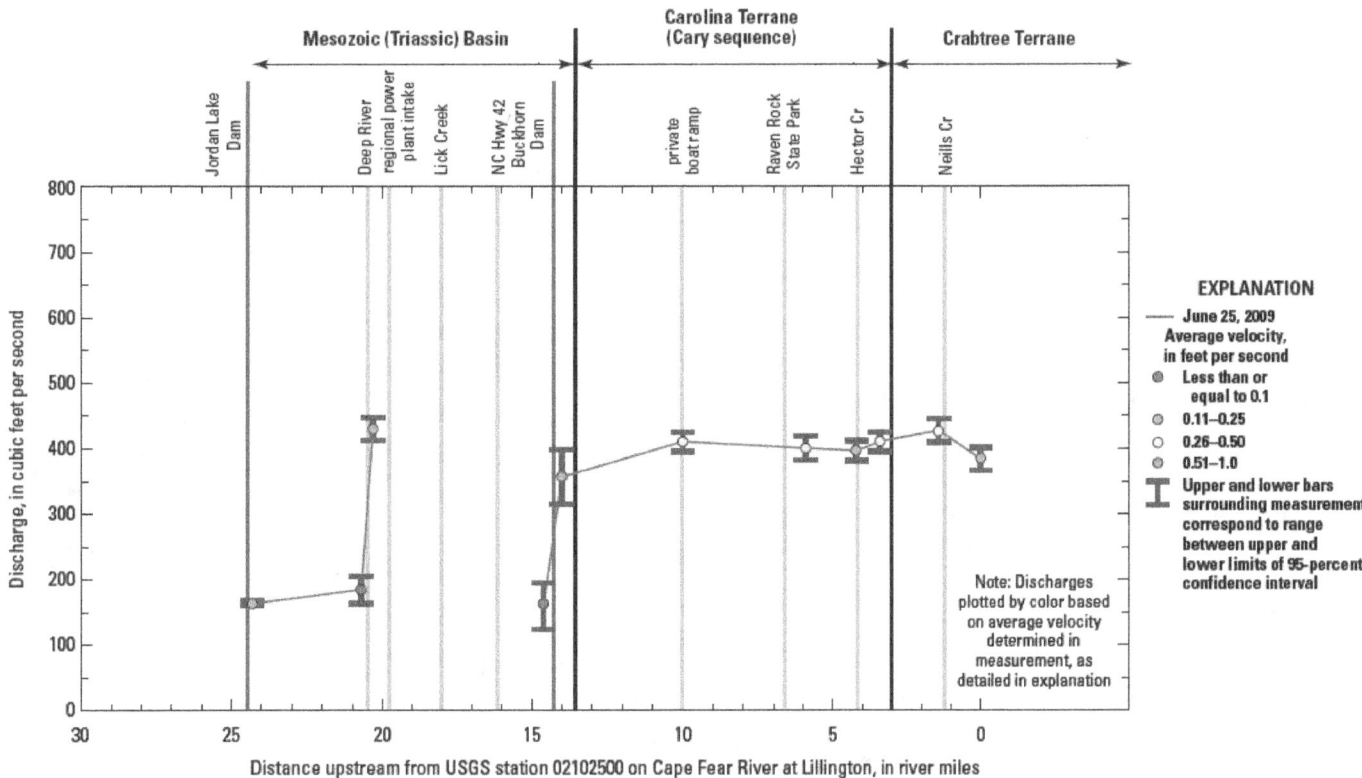

Figure 15. Measured discharges on June 25, 2009, on the Haw and Cape Fear Rivers between Jordan Lake Dam and Lillington, North Carolina.

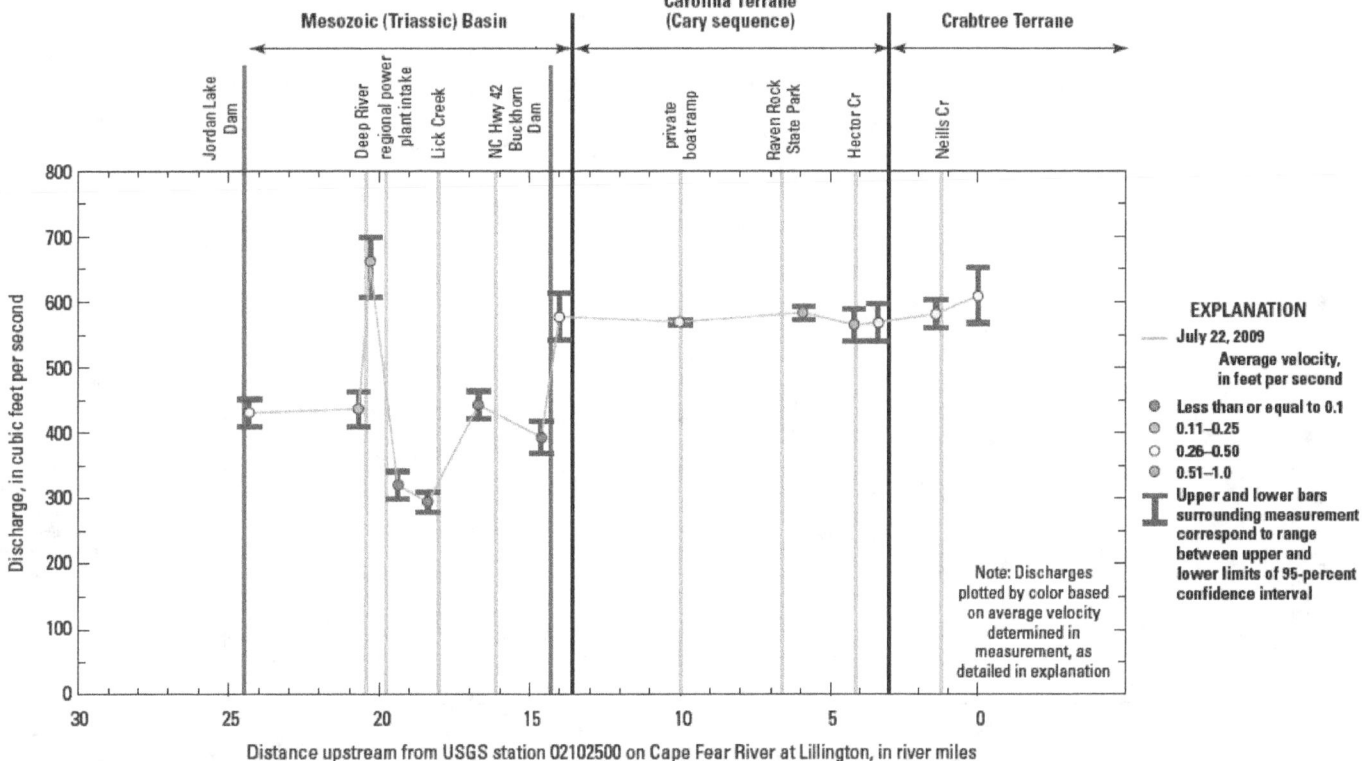

Figure 16. Measured discharges on July 22, 2009, on the Haw and Cape Fear Rivers between Jordan Lake Dam and Lillington, North Carolina.

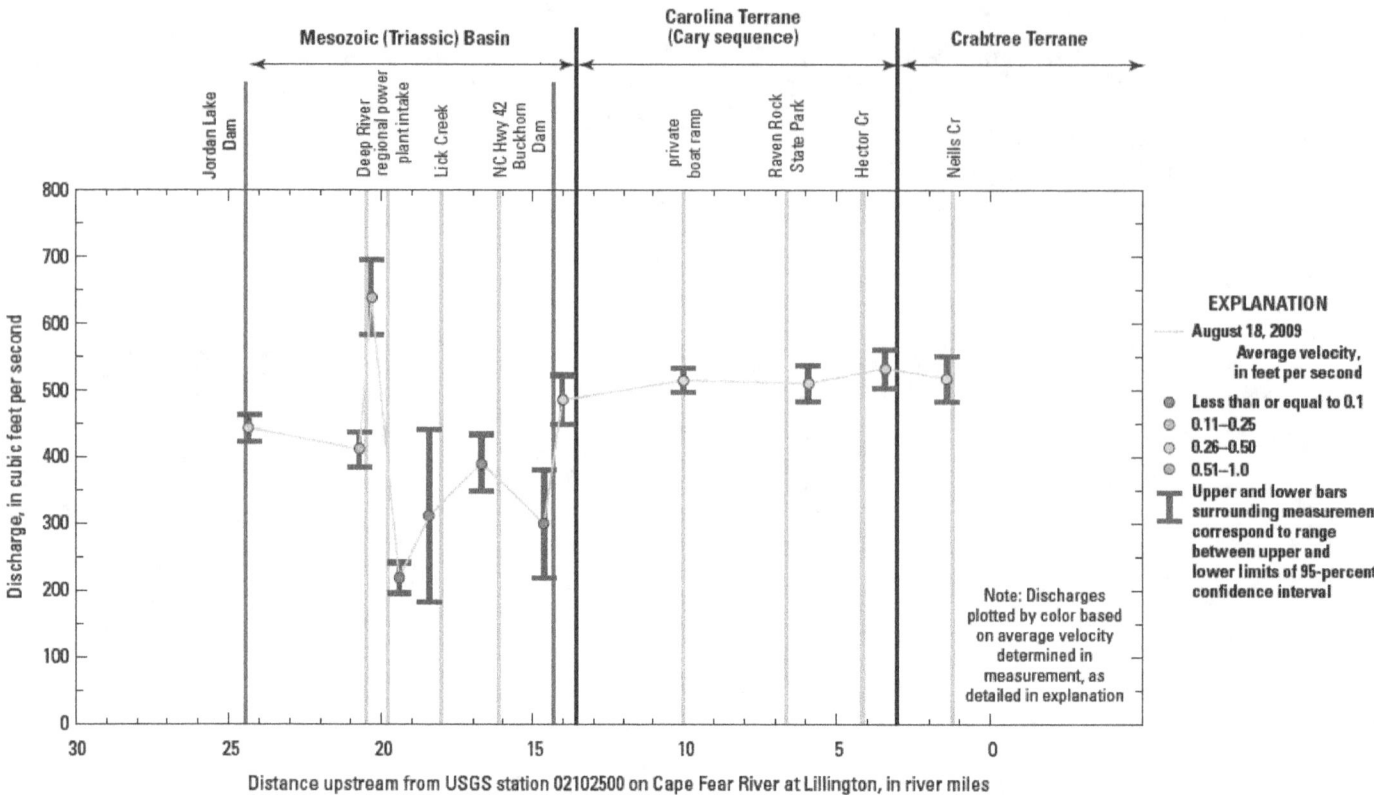

Figure 17. Measured discharges on August 18, 2009, on the Haw and Cape Fear Rivers between Jordan Lake Dam and Lillington, North Carolina.

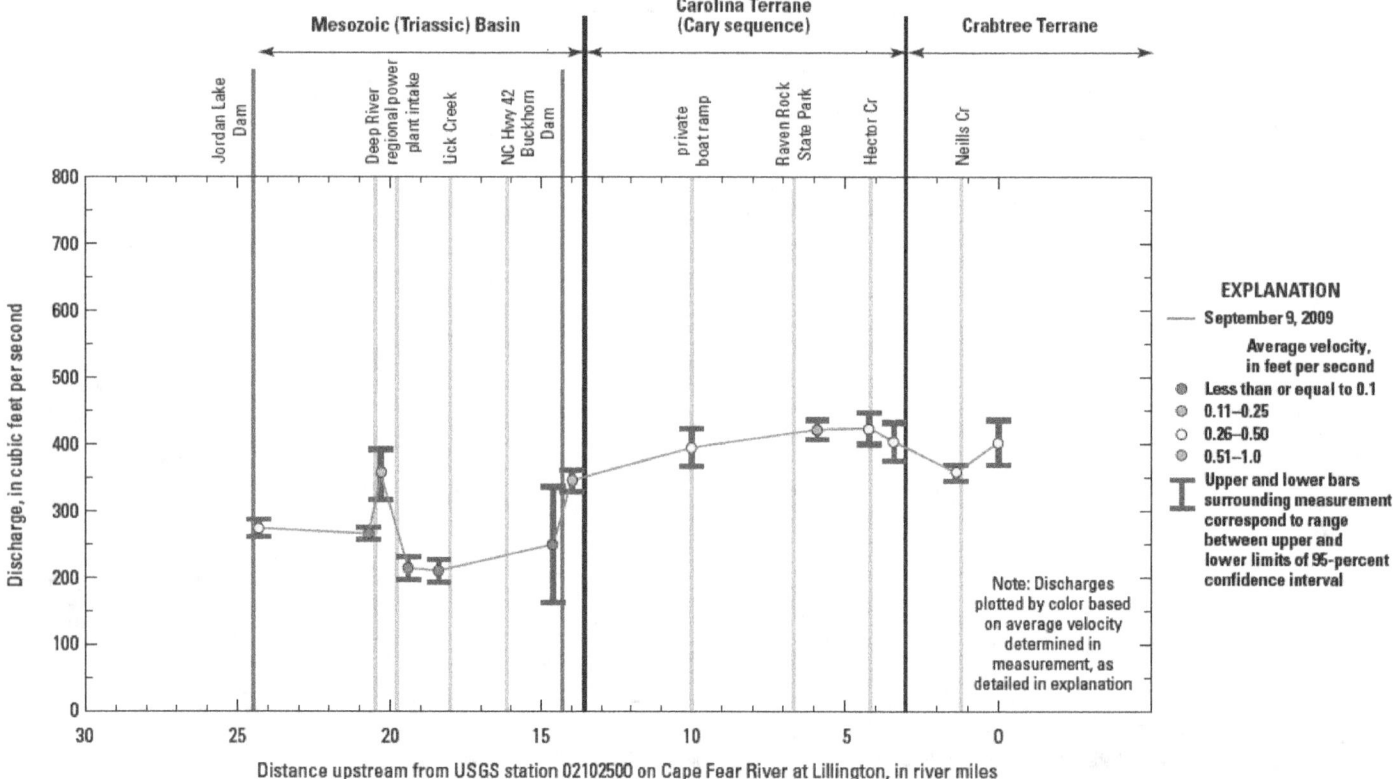

Figure 18. Measured discharges on September 9, 2009, on the Haw and Cape Fear Rivers between Jordan Lake Dam and Lillington, North Carolina.

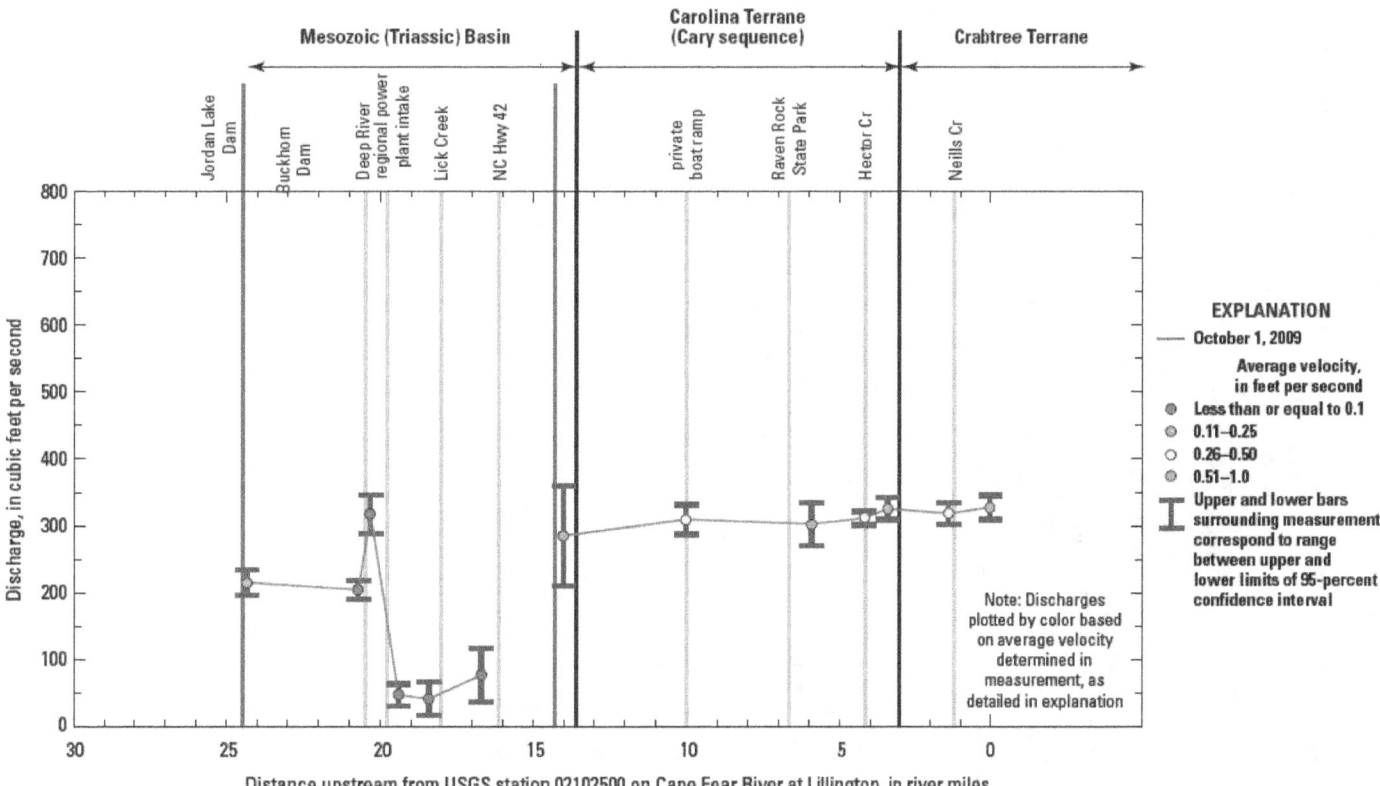

Figure 19. Measured discharges on October 1, 2009, on the Haw and Cape Fear Rivers between Jordan Lake Dam and Lillington, North Carolina.

acoustic discharge measurements with 95-percent confidence intervals exceeding 8 percent are rated as poor. The flow dynamics associated with the confluence of two large basins (Deep and Haw Rivers) in a backwater reach during base-flow conditions on five of the six series of measurements were reflected in the quality rating of the measured discharges at these locations.

Comparisons of the differences in measured discharges between sites 5 and 6 on the Cape Fear River upstream and downstream, respectively, from the intake at site F7 (fig. 6) with the reported water withdrawals also highlighted challenges during the synoptic measurements. Differences in measured discharges between sites 5 and 6 on April 27, July 22, August 18, September 9, and October 1 were 60, 83, 106, 66, and 68 percent, respectively, of the reported water withdrawals. For five of the six synoptic measurements, the reported water withdrawals were identical – about 396 ft³/s. As discussed in a subsequent section, reported water withdrawals at site F7 were questioned when compared to the discharge records at the USGS streamgage on the adjacent diversion canal downstream from the regional power plant (fig. 6, site 8). Comparisons between the reported withdrawals and the differences in measured discharges at sites 5 and 6 were, therefore, limited because of the questions concerning the accuracy of the withdrawal data.

Several noticeable patterns can be seen in the graphs of each of the series of synoptic measurements (figs. 14–19) when inspected from an overall "big picture" perspective.

The largest water diversion occurs just downstream from the confluence of the Haw and Deep Rivers where water is withdrawn for cooling purposes at the regional power plant (fig. 6, site F7). Discharges measured during the synoptic measurements decreased substantially in the main stem downstream from the power-plant intake point. Downstream from Buckhorn Dam, little gain or loss is evident between the dam and Raven Rock State Park (fig. 7, site 22), although minor fluctuations in flow patterns occur between the State park and the Lillington streamgage (fig. 7, site 28). What is clear in the graphs is that flows in the Cape Fear River between the intake point below the confluence and Buckhorn Dam are governed by the water diversion through the regional power plant. Upon release from the plant, the flow travels approximately 6.3 mi along a diversion canal that drains into the main stem immediately upstream from Buckhorn Dam.

Following the synoptic measurements on April 27, 2009, a breach was discovered between the diversion canal and Cape Fear River near the northern end of McKay Island (fig. 6). Thus, flow was measured in the unnamed stream (fig. 6, site 11) between the canal and river during the remaining five synoptic measurements. The measured discharges in the unnamed stream ranged from 43 to 52 percent of the measured discharge at the next upstream canal site (site 9). Although a few other smaller breaches between the canal and Cape Fear River were noted during the field operations, none was observed to be as large as the breach near McKay Island.

Table 4. Results of synoptic discharge measurements made on April 27, 2009, on the Haw and Cape Fear Rivers between Jordan Lake Dam and Lillington, North Carolina.

[USGS, U.S. Geological Survey; main stem refers to the Haw and Cape Fear Rivers; ft³/s, cubic foot per second; change, in percent, computed as difference in measured discharge from next upstream measured discharge on main stem; DA, doesn't apply; NA, not available; --, not measured; measurement rating determined from base uncertainty at 95-percent confidence interval: (F), fair; (G), good; (P), poor (http://nc.water.usgs.gov/usgs/info/qaplan/surface.html)]

Site index number (figs. 1, 6, 7)	USGS station number	Station name	Main stem river miles upstream from Lillington streamgage (site 28)	Measured discharge, in ft³/s	Average velocity, in ft/s	Standard deviation, in ft/s	95-percent confidence interval of measurement uncertainty, in percent (measurement rating)	Lowest discharge in 95-percent confidence interval, in ft³/s	Highest discharge in 95-percent confidence interval, in ft³/s	Change, in percent
					Sites above Buckhorn Dam					
1	02098198	Haw R below B. Everett Jordan Dam near Moncure, NC	24.3	859	0.91	29.3	5.5 (F)	812	906	DA
F1		Industrial	NA	Withdrawal: 0.24 ft³/s				Return: 0.1 ft³/s		
F2		Industrial	NA	Withdrawal: Not recorded				Return: Not recorded		
F3		Industrial	NA	Withdrawal: Not recorded				Return: 0.08 ft³/s		
F4		Industrial	NA	Withdrawal: 0.03 ft³/s				Return: 0 ft³/s		
2	02098210	Haw River above Deep River near Haywood, NC	20.7	835	0.34	31.3	6.0 (F)	785	885	-2.8
3	02102000	Deep River at Moncure, NC	DA	448 [1]	--	--	--	--	--	DA
F6		Hydropower power generation	NA	Power generation online all day.						
4	0210204915	Deep River above mouth near Moncure, NC	DA	432	0.18	34.9	6.5 (F)	404	460	DA
5	02102050	Cape Fear River above powerplant intake near Moncure, NC	20.3	1,200	0.35	18.8	2.5 (G)	1,171	1,231	43.8
F7		Regional power utility	NA	Withdrawal: 395.77 ft³/s				Return: 338.68 ft³/s		
6	02102082	Cape Fear River below powerplant intake near Moncure, NC	19.4	960	0.22	22.9	3.8 (G)	923	997	-20.1
7	02102090	Cape Fear River above railroad bridge near Rosser, NC	18.4	--	--	--	--	--	--	DA
8	02102094	Cape Fear powerplant discharge canal near Brickhaven, NC	DA	339 [1]	--	--	--	--	--	DA
F8		Industrial	NA	Withdrawal: 0 ft³/s				Return: No return discharge		
9	0210209450	Cape Fear powerplant discharge canal above juncture at Brickhaven, NC	DA	--	--	--	--	--	--	DA
10	0210209475	Cape Fear powerplant discharge canal below dam at Brickhaven, NC	DA	--	--	--	--	--	--	DA
11	0210209625	Unnamed stream between canal and Cape Fear River at Brickhaven, NC	DA	--	--	--	--	--	--	DA
12	0210209650	Cape Fear powerplant discharge canal below juncture at Brickhaven, NC	DA	--	--	--	--	--	--	DA
13	02102159	Lick Creek near Rosser, NC	DA	5.26	0.105	--	--	--	--	DA

Table 4. Results of synoptic discharge measurements made on April 27, 2009, on the Haw and Cape Fear Rivers between Jordan Lake Dam and Lillington, North Carolina.—Continued

[USGS, U.S. Geological Survey; main stem refers to the Haw and Cape Fear Rivers; ft³/s, cubic foot per second; change, in percent, computed as difference in measured discharge from next upstream measured discharge on main stem; DA, doesn't apply; NA, not available; --, not measured; measurement rating determined from base uncertainty at 95-percent confidence interval: (F), fair; (G), good; (P), poor (http://nc.water.usgs.gov/usgs/info/qaplan/surface.html)]

Site index number (figs. 1, 6, 7)	USGS station number	Station name	Main stem river miles upstream from Lillington streamgage (site 28)	Measured discharge, in ft³/s	Average velocity, in ft/s	Standard deviation, in ft³/s	95-percent confidence interval of measurement uncertainty, in percent (measurement rating)	Lowest discharge in 95-percent confidence interval, in ft³/s	Highest discharge in 95-percent confidence interval, in ft³/s	Change, in percent
14	0210215960	Cape Fear River above NC Highway 42 near Corinth, NC	16.7	1,140	0.21	41.1	5.8 (F)	1,076	1,208	DA
F9		Municipal	NA	Withdrawal: 10.84 ft³/s				Return: 0.59 ft³/s [2]		DA
15	0210215995	Cape Fear powerplant discharge canal at NC Highway 42 near Corinth, NC	DA	--	--	--	--	--	--	DA
16	02102176	Cape Fear powerplant discharge canal above mouth near Corinth, NC	DA	--	--	--	--	--	--	DA
17	02102177	Cape Fear River upstream from Buckhorn Dam near Corinth, NC	14.6	1,210	0.16	26.5	3.5 (G)	1,168	1,252	6.0
Sites at or below Buckhorn Dam										
18	02102178	Cape Fear River at Buckhorn Dam near Corinth, NC	14.3							
19	0210217810	Cape Fear River downstream from Buckhorn Dam near Corinth, NC	14	1,230	0.53	27.9	3.6 (G)	1,189	1,279	2.0
20	02102192	Buckhorn Creek near Corinth, NC	DA	25.9	0.58	--	--	--	--	DA
21	02102240	Cape Fear River adjacent to Bay Street near Cokesbury, NC	10.	1,330	0.79	14.5	1.7 (G)	1,304	1,350	7.5
22	02102265	Cape Fear River at Raven Rock State Park group camp, NC	5.9	1,270	0.91	35	4.4 (G)	1,217	1,329	-4.1
23	02102278	Cape Fear River above Hector Creek near Chalybeate, NC	4.2	1,290	1.03	55.2	6.8 (F)	1,206	1,382	1.6
24	02102280	Hector Creek near Chalybeate, NC	DA	6.31	0.729	--	--	--	--	DA
25	02102283	Cape Fear R adjacent to Bradley Road near Lillington, NC	3.4	1,360	0.71	37.3	4.4 (G)	1,302	1,422	5.3
26	02102289	Cape Fear River above Neills Creek near Lillington, NC	1.4	1,340	0.69	30.4	3.6 (G)	1,295	1,393	-1.3
27	02102480	Neills Creek near Lillington, NC	DA	11.4	0.164	--	--	--	--	DA
F10		Municipal	NA	Withdrawal: 23.73 ft³/s				Return: Outside of study area		DA
28	02102500	Cape Fear River at Lillington, NC	0.0	1,400	0.7	46.3	5.3 (F)	1,321	1,469	3.8

[1]Computed DAily discharge at this site for indicated DAte.

[2]Reported return discharge associated with water-treatment operations that discharge to a tributary to the Deep River upstream from the study area.

Table 5. Results of synoptic discharge measurements made on June 25, 2009, on the Haw and Cape Fear Rivers between Jordan Lake Dam and Lillington, North Carolina.

[USGS, U.S. Geological Survey; main stem refers to the Haw and Cape Fear Rivers; ft³/s, cubic feet per second; change, in percent, computed as difference in measured discharge from next upstream measured discharge on main stem; DA, doesn't apply; NA, not available; --, not measured; measurement rating determined from base uncertainty at 95-percent confidence interval: (G), good, (P), poor, (F), fair (http://nc.water.usgs.gov/usgs/info/qaplan/surface.html)]

Site index number (figs. 1, 6, 7)	USGS station number	Station name	Main stem river miles upstream from Lillington streamgage (site 28)	Measured discharge, in ft³/s	Average velocity, in ft/s	Standard deviation, in ft³/s	95-percent confidence interval of measurement uncertainty, in percent (measurement rating)	Lowest discharge in 95-percent confidence interval, in ft³/s	Highest discharge in 95-percent confidence interval, in ft³/s	Change, in percent
						Measured discharge				
			Sites above Buckhorn Dam							
1	02098198	Haw R below B. Everett Jordan Dam near Moncure, NC	24.3	164	0.16	5.5	2.7 (G)	160	168	DA
F1		Industrial	NA	Withdrawal: 0.36 ft³/s				Return: 0.11 ft³/s		
F2		Industrial	NA	Withdrawal: 0 ft³/s				Return: Not recorded		
F3		Industrial	NA	Withdrawal: 0.36 ft³/s				Return: 0.02 ft³/s		
F4		Industrial	NA	Withdrawal: 0.15 ft³/s				Return: 0 ft³/s		
2	02098210	Haw River above Deep River near Haywood, NC	20.7	185	0.06	20.2	11.5 (P)	164	206	12.8
3	02102000	Deep River at Moncure, NC	DA	236 [1]	--	--	--	--	--	DA
F6		Hydropower power generation	NA	Power generation offline all day.						
4	0210204915	Deep River above mouth near Moncure, NC	DA	262	0.11	27.1	16.5 (P)	219	305	DA
5	02102050	Cape Fear River above powerplant intake near Moncure, NC	20.3	431	0.15	10.4	3.9 (G)	414	448	133.0
F7		Regional power utility	NA	Withdrawal: 395.77 ft³/s				Return: 329.4 ft³/s		
6	02102082	Cape Fear River below powerplant intake near Moncure, NC	19.4	--	--	--	--	--	--	--
7	02102090	Cape Fear River above railroad bridge near Rosser, NC	18.4	--	--	--	--	--	--	--
8	02102094	Cape Fear powerplant discharge canal near Brickhaven, NC	DA	325	1.05	10.8	5.3 (F)	308	342	DA
F8		Industrial	NA	Withdrawal: 0 ft³/s				Return: No return discharge		
9	0210209450	Cape Fear powerplant discharge canal above juncture at Brickhaven, NC	DA	332	1.33	2.1	1.0 (G)	329	335	DA
10	0210209475	Cape Fear powerplant discharge canal below dam at Brickhaven, NC	DA	0.034	0.008	--	--	--	--	DA
11	0210209625	Unnamed stream between canal and Cape Fear River at Brickhaven, NC	DA	149	1.45	7.4	4.0 (G)	143	155	DA
12	0210209650	Cape Fear powerplant discharge canal below juncture at Brickhaven, NC	DA	178	0.97	4.1	3.7 (G)	171	185	DA
13	02102159	Lick Creek near Rosser, NC	DA	.33	0.008	--	--	--	--	DA

Table 5.   Results of synoptic discharge measurements made on June 25, 2009, on the Haw and Cape Fear Rivers between Jordan Lake Dam and Lillington, North Carolina.—Continued

[USGS, U.S. Geological Survey; main stem refers to the Haw and Cape Fear Rivers; ft³/s, cubic feet per second; change, in percent, computed as difference in measured discharge from next upstream measured discharge on main stem; DA, doesn't apply; NA, not available; --, not measured; measurement rating determined from base uncertainty at 95-percent confidence interval: (G), good, (P), poor, (F), fair (http://nc.water.usgs.gov/usgs/info/qaplan/surface.html)]

Site index number (figs. 1, 6, 7)	USGS station number	Station name	Main stem river miles upstream from Lillington streamgage (site 28)	Measured discharge						Change, in percent
				Measured discharge, in ft³/s	Average velocity, in ft/s	Standard deviation, in ft³/s	95-percent confidence interval of measurement uncertainty, in percent (measurement rating)	Lowest discharge in 95-percent confidence interval, in ft³/s	Highest discharge in 95-percent confidence interval, in ft³/s	
14	0210215960	Cape Fear River above NC Highway 42 near Corinth, NC	16.7	--	--	--	--	--	--	--
F9		Municipal	NA	Withdrawal: 12.48 ft³/s				Return: 0.59 ft³/s [2]		
15	0210215995	Cape Fear powerplant discharge canal at NC Highway 42 near Corinth, NC	DA	207	0.79	14.6	5.6 (F)	195	219	DA
16	02102176	Cape Fear powerplant discharge canal above mouth near Corinth, NC	DA	190	0.62	5.9	5.0 (G)	181	199	DA
17	02102177	Cape Fear River upstream from Buckhorn Dam near Corinth, NC	14.6	160	0.02	22.	22.0 (P)	125	195	DA
		Sites at or below Buckhorn Dam								
18	02102178	Cape Fear River at Buckhorn Dam near Corinth, NC	14.3							
19	0210217810	Cape Fear River downstream from Buckhorn Dam near Corinth, NC	14	357	0.22	52.6	11.8 (P)	315.	399	123.1
20	02102192	Buckhorn Creek near Corinth, NC	DA	0.0	--	--	--	--	--	DA
21	02102240	Cape Fear River adjacent to Bay Street near Cokesbury, NC	10	411	0.333	9.6	3.7 (G)	396	426	15.1
22	02102265	Cape Fear River at Raven Rock State Park group camp, NC	5.9	401	0.497	22.5	4.5 (G)	383	419	-2.4
23	02102278	Cape Fear River above Hector Creek near Chalybeate, NC	4.2	397	0.519	18.3	3.7 (G)	382	412	-1.0
24	02102280	Hector Creek near Chalybeate, NC	DA	1.17	0.205	--	--	--	--	DA
25	02102283	Cape Fear R adjacent to Bradley Road near Lillington, NC	3.4	410	0.313	18.5	3.6 (G)	396	426	3.5
26	02102289	Cape Fear River above Neills Creek near Lillington, NC	1.4	428	0.338	11.5	4.3 (G)	410	446	4.1
27	02102480	Neills Creek near Lillington, NC	DA	1.61	0.024	--	--	--	--	DA
F10		Municipal	NA	Withdrawal: 25.39 ft³/s				Return: Outside of study area		
28	02102500	Cape Fear River at Lillington, NC	0.0	384	0.19	22	4.6 (G)	366	402	-10.3

[1] Computed daily discharge at this site for indicated date.

[2] Reported return discharge associated with water-treatment operations, not the wastewater-treatment plant operations that discharge to a tributary to the Deep River upstream from the study area.

Table 6.   Results of synoptic discharge measurements made on July 22, 2009, on the Haw and Cape Fear Rivers between Jordan Lake Dam and Lillington, North Carolina.

[USGS, U.S. Geological Survey; main stem refers to the Haw and Cape Fear Rivers; ft³/s, cubic foot per second; ft/s, foot per second; change, in percent, computed as difference in measured discharge from next upstream measured discharge on main stem; DA, doesn't apply; NA, not available; --, not measured; measurement rating determined from base uncertainty at 95-percent confidence interval: (G), good; (F) fair; (P), poor (http://nc.water.usgs.gov/usgs/info/qaplan/surface.html)]

Site index number (figs. 1, 6, 7)	USGS station number	Station name	Main stem river miles upstream from Lillington streamgage (site 28)	Measured discharge, in ft³/s	Average velocity, in ft/s	Standard deviation, in ft³/s	95-percent confidence interval of measurement uncertainty, in percent (measurement rating)	Lowest discharge in 95-percent confidence interval, in ft³/s	Highest discharge in 95-percent confidence interval, in ft³/s	Change, in percent
Sites above Buckhorn Dam										
1	02098198	Haw R below B. Everett Jordan Dam near Moncure, NC	24.3	430	0.38	13.2	4.9 (G)	409	451	DA
F1		Industrial	NA	Withdrawal: 0.33 ft³/s				Return: 0.1 ft³/s		
F2		Industrial	NA	Withdrawal: 0 ft³/s				Return: 0 ft³/s		
F3		Industrial	NA	Withdrawal: Not recorded				Return: 0.06 ft³/s		
F4		Industrial	NA	Withdrawal: 0.11 ft³/s				Return: 0 ft³/s		
2	02098210	Haw River above Deep River near Haywood, NC	20.7	436	0.17	16	5.9 (F)	410	462	1.4
3	02102000	Deep River at Moncure, NC	DA	164[1]	--	--	--	--	--	DA
F6		Hydropower power generation	NA	Power generation online between midnight and 03:00, then between 13:00 and 16:00; offline during remainder of day.						
4	0210204915	Deep River above mouth near Moncure, NC	DA	193	0.09	15.3	6.3 (F)	181	205	DA
5	02102050	Cape Fear River above powerplant intake near Moncure, NC	20.3	653	0.22	28.9	7.1 (F)	607	699	49.8
F7		Regional power utility	NA	Withdrawal: 395.77 ft³/s				Return: 338.53 ft³/s		
6	02102082	Cape Fear River below powerplant intake near Moncure, NC	19.4	320	0.09	13	6.5 (F)	299	341	-51.0
7	02102090	Cape Fear River above railroad bridge near Rosser, NC	18.4	294	0.067	10	5.4 (F)	278	310	-8.1
8	02102094	Cape Fear powerplant discharge canal near Brickhaven, NC	DA	355	1.02	3.9	1.8 (G)	349	361	DA
F8		Industrial	NA	Withdrawal: 0 ft³/s				Return: No return discharge		
9	0210209450	Cape Fear powerplant discharge canal above juncture at Brickhaven, NC	DA	319	1.34	11.1	3.7 (G)	307	331	DA
10	0210209475	Cape Fear powerplant discharge canal below dam at Brickhaven, NC	DA	--	--	--	--	--	--	DA
11	0210209625	Unnamed stream between canal and Cape Fear River at Brickhaven, NC	DA	160.	1.5	3.7	3.7 (G)	154	166	DA
12	0210209650	Cape Fear powerplant discharge canal below juncture at Brickhaven, NC	DA	202	1.06	4.8	1.9 (G)	198	206	DA

Table 6. Results of synoptic discharge measurements made on July 22, 2009, on the Haw and Cape Fear Rivers between Jordan Lake Dam and Lillington, North Carolina.—Continued

[USGS, U.S. Geological Survey; main stem refers to the Haw and Cape Fear Rivers; ft³/s, cubic foot per second; change, in percent, computed as difference in measured discharge from next upstream measured discharge on main stem; DA, doesn't apply; NA, not available; --, not measured; measurement rating determined from base uncertainty at 95-percent confidence interval: (G), good; (F) fair; (P), poor (http://nc.water.usgs.gov/usgs/info/qaplan/surface.html)]

Site index number (figs. 1, 6, 7)	USGS station number	Station name	Main stem river miles upstream from Lillington streamgage (site 28)	Measured discharge, in ft³/s	Average velocity, in ft/s	Standard deviation, in ft³/s	95-percent confidence interval of measurement uncertainty, in percent (measurement rating)	Lowest discharge in 95-percent confidence interval, in ft³/s	Highest discharge in 95-percent confidence interval, in ft³/s	Change, in percent
13	02102159	Lick Creek near Rosser, NC	DA	--	--	--	--	--	--	DA
14	0210215960	Cape Fear River above NC Highway 42 near Corinth, NC	16.7	442	0.08	13.3	4.8 (G)	421	463	50.3
F9	Municipal		NA	Withdrawal: 11.77 ft³/s				Return: 1.01 ft³/s [2]		
15	0210215995	Cape Fear powerplant discharge canal at NC Highway 42 near Corinth, NC	DA	198	0.785	1.8	1.4 (G)	195	201	DA
16	02102176	Cape Fear powerplant discharge canal above mouth near Corinth, NC	DA	194	0.604	7.4	6.1 (F)	182	206	DA
17	02102177	Cape Fear River upstream from Buckhorn Dam near Corinth, NC	14.6	393	0.06	31.2	6.4 (F)	368	418	-11.1
		Sites at or below Buckhorn Dam								
18	02102178	Cape Fear River at Buckhorn Dam near Corinth, NC	14.3							
19	0210217810	Cape Fear River downstream from Buckhorn Dam near Corinth, NC	14	577	0.312	22.6	6.3 (F)	541	613	46.8
20	02102192	Buckhorn Creek near Corinth, NC	DA	0.57	0.02	--	--	--	--	DA
21	02102240	Cape Fear River adjacent to Bay Street near Cokesbury, NC	10	569	0.437	2.4	0.7 (G)	565	573	-1.4
22	02102265	Cape Fear River at Raven Rock State Park group camp, NC	5.9	583	0.669	6.2	1.7 (G)	573	593	2.5
23	02102278	Cape Fear River above Hector Creek near Chalybeate, NC	4.2	564	0.536	30.2	4.3 (G)	540	588	-3.3
24	02102280	Hector Creek near Chalybeate, NC	DA	0.84	0.171	--	--	--	--	DA
25	02102283	Cape Fear R adjacent to Bradley Road near Lillington, NC	3.4	568	0.374	36	5.1 (F)	539	597	0.7
26	02102289	Cape Fear River above Neills Creek near Lillington, NC	1.4	581	0.431	13	3.6 (G)	560	602	2.3
27	02102480	Neills Creek near Lillington, NC	DA	1.18	0.019	--	--	--	--	DA
F10	Municipal		NA	Withdrawal: 20.27 ft³/s				Return: Outside of study area		
28	02102500	Cape Fear River at Lillington, NC	0.0	609	0.39	26	6.8 (F)	567	651	4.8

[1]Computed daily discharge at this site for indicated date.

[2]Reported return discharge associated with water-treatment operations, not the wastewater-treatment plant operations, that discharge to a tributary to the Deep River upstream from the study area.

Table 7. Results of synoptic discharge measurements made on August 18, 2009, on the Haw and Cape Fear Rivers between Jordan Lake Dam and Lillington, North Carolina.

[USGS, U.S. Geological Survey; main stem refers to the Haw and Cape Fear Rivers; ft³/s, cubic foot per second; change, in percent, computed as difference in measured discharge from next upstream measured discharge on main stem; DA, doesn't apply; NA, not available; --, not measured; measurement rating determined from base uncertainty at 95-percent confidence interval: (G), good; (F), fair; (P), poor (http://nc.water.usgs.gov/usgs/info/qaplan/surface.html)]

Site index number (figs. 1, 6, 7)	USGS station number	Station name	Main stem river miles upstream from Lillington streamgage (site 28)	Measured discharge, in ft³/s	Average velocity, in ft/s	Standard deviation, in ft³/s	95-percent confidence interval of measurement uncertainty, in percent (measurement rating)	Lowest discharge in 95-percent confidence interval, in ft³/s	Highest discharge in 95-percent confidence interval, in ft³/s	Change, in percent
							Measured discharge			
		Sites above Buckhorn Dam								
1	02098198	Haw R below B. Everett Jordan Dam near Moncure, NC	24.3	443	0.41	12.6	4.6 (G)	423	463	DA
F1		Industrial	NA	Withdrawal: 0.33 ft³/s				Return: 0.09 ft³/s		
F2		Industrial	NA	Withdrawal: Not recorded				Return: 0 ft³/s		
F3		Industrial	NA	Withdrawal: 0.31 ft³/s				Return: 0.1 ft³/s		
F4		Industrial	NA	Withdrawal: 0.2 ft³/s				Return: 0 ft³/s		
2	02098210	Haw River above Deep River near Haywood, NC	20.7	411	0.15	16.2	6.3 (F)	385	437	-7.2
3	02102000	Deep River at Moncure, NC	DA	90.3	0.07	10.8	9.6 (P)	82	99	DA
F6		Hydropower power generation	NA	Power generation offline all day.						
4	0210204915	Deep River above mouth near Moncure, NC	DA	101	0.05	24.2	19.2 (P)	82	120	DA
5	02102050	Cape Fear River above powerplant intake near Moncure, NC	20.3	639	0.19	34.8	8.7 (P)	583	695	55.5
F7		Regional power utility	NA	Withdrawal: 395.77 ft³/s				Return: 337.29 ft³/s		
6	02102082	Cape Fear River below powerplant intake near Moncure, NC	19.4	219	0.06	14.1	10.3 (P)	196	242	-65.7
7	02102090	Cape Fear River above railroad bridge near Rosser, NC	18.4	312	0.067	80.4	41.2 (P)	183	441	42.5
8	02102094	Cape Fear powerplant discharge canal near Brickhaven, NC	DA	351	1.02	2.1	1.0 (G)	348	354	DA
F8		Industrial	NA	Withdrawal: 0 ft³/s				Return: No return discharge		
9	0210209450	Cape Fear powerplant discharge canal above juncture at Brickhaven, NC	DA	341	1.33	2.6	1.2 (G)	337	345	DA
10	0210209475	Cape Fear powerplant discharge canal below dam at Brickhaven, NC	DA	--	--	--	--	--	--	DA
11	0210209625	Unnamed stream between canal and Cape Fear River at Brickhaven, NC	DA	157	1.43	3.2	3.2 (G)	152	162	DA
12	0210209650	Cape Fear powerplant discharge canal below juncture at Brickhaven, NC	DA	182	0.95	0.4	0.3 (G)	181	183	DA
13	02102159	Lick Creek near Rosser, NC	DA	0.18	0.011	--	--	--	--	DA

Table 7. Results of synoptic discharge measurements made on August 18, 2009, on the Haw and Cape Fear Rivers between Jordan Lake Dam and Lillington, North Carolina.—Continued

[USGS, U.S. Geological Survey; main stem refers to the Haw and Cape Fear Rivers; ft³/s, cubic foot per second; change, in percent, computed as difference in measured discharge from next upstream measured discharge on main stem; DA, doesn't apply; NA, not available; --, not measured; measurement rating determined from base uncertainty at 95-percent confidence interval: (G), good; (F), fair; (P), poor (http://nc.water.usgs.gov/usgs/info/qaplan/surface.html)]

Site index number (figs. 1, 6, 7)	USGS station number	Station name	Main stem river miles upstream from Lillington streamgage (site 28)	Measured discharge, in ft³/s	Average velocity, in ft/s	Standard deviation, in ft³/s	95-percent confidence interval of measurement uncertainty, in percent (measurement rating)	Lowest discharge in 95-percent confidence interval, in ft³/s	Highest discharge in 95-percent confidence interval, in ft³/s	Change, in percent
14	0210215960	Cape Fear River above NC Highway 42 near Corinth, NC	16.7	391	0.08	26.8	11.0 (P)	348	434	25.3
F9		Municipal	NA	Withdrawal: 10.27 ft³/s				Return: 0.53 ft³/s [1]		
15	0210215995	Cape Fear powerplant discharge canal at NC Highway 42 near Corinth, NC	DA	182	0.68	4.9	4.3 (G)	174	190	DA
16	02102176	Cape Fear powerplant discharge canal above mouth near Corinth, NC	DA	184	0.6	5.3	4.6 (G)	176	192	DA
17	02102177	Cape Fear River upstream from Buckhorn Dam near Corinth, NC	14.6	300	0.04	101	26.9 (P)	219	381	-23.3
		Sites at or below Buckhorn Dam								
18	02102178	Cape Fear River at Buckhorn Dam near Corinth, NC	14.3							
19	0210217810	Cape Fear River downstream from Buckhorn Dam near Corinth, NC	14	485	0.26	23.1	7.6 (F)	448	522.	61.7
20	02102192	Buckhorn Creek near Corinth, NC	DA	0.0	--	--	--	--	--	DA
21	02102240	Cape Fear River adjacent to Bay Street near Cokesbury, NC	10	514	0.403	11.5	3.6 (G)	496	532	6.0
22	02102265	Cape Fear River at Raven Rock State Park group camp, NC	5.9	510	0.498	16.8	5.3 (F)	483	537	-0.8
23	02102278	Cape Fear River above Hector Creek near Chalybeate, NC	4.2	--	--	--	--	--	--	DA
24	02102280	Hector Creek near Chalybeate, NC	DA	0.81	0.225	--	--	--	--	DA
25	02102283	Cape Fear R adjacent to Bradley Road near Lillington, NC	3.4	532	0.369	18.1	5.4 (F)	503	561	DA
26	02102289	Cape Fear River above Neills Creek near Lillington, NC	1.4	517	0.382	21.5	6.7 (F)	483	551	-2.8
27	02102480	Neills Creek near Lillington, NC	DA	1.78	0.027	--	--	--	--	DA
F10		Municipal	NA	Withdrawal: 20.7 ft³/s				Return: Outside of study area		
28	02102500	Cape Fear River at Lillington, NC	0.0	512 [2]	--	--	--	--	--	DA

[1] Reported return discharge associated with water-treatment plant operations, not the wastewater-treatment plant operations, that discharge to a tributary to the Deep River upstream from the study area.

[2] Computed daily discharge at this site for indicated date.

Table 8. Results of synoptic discharge measurements made on September 9, 2009, on the Haw and Cape Fear Rivers between Jordan Lake Dam and Lillington, North Carolina.

[USGS, U.S. Geological Survey; main stem refers to the Haw and Cape Fear Rivers; ft^3/s, cubic foot per second; ft/s, foot per second; change, in percent, computed as difference in measured discharge from next upstream measured discharge on main stem; DA, doesn't apply; NA, not available; --, not measured; measurement rating determined from base uncertainty at 95-percent confidence interval: (G), good; (F), fair, (P), poor (http://nc.water.usgs.gov/usgs/info/qaplan/surface.html)]

Site index number (figs. 1, 6, 7)	USGS station number	Station name	Main stem river miles upstream from Lillington streamgage (site 28)	Measured discharge, in ft^3/s	Average velocity, in ft/s	Standard deviation, in ft^3/s	95-percent confidence interval of measurement uncertainty, in percent (measurement rating)	Measured discharge		
								Lowest discharge in 95-percent confidence interval, in ft^3/s	Highest discharge in 95-percent confidence interval, in ft^3/s	Change, in percent
Sites above Buckhorn Dam										
1	02098198	Haw R below B. Everett Jordan Dam near Moncure, NC	24.3	274	0.29	7.9	4.6 (G)	261	287	DA
F1		Industrial	NA	Withdrawal: 0.35 ft^3/s				Return: 0.07 ft^3/s		
F2		Industrial	NA	Withdrawal: 0 ft^3/s				Return: 0 ft^3/s		
F3		Industrial	NA	Withdrawal: 0.17 ft^3/s				Return: 0.05 ft^3/s		
F4		Industrial	NA	Withdrawal: 0.09 ft^3/s				Return: 0 ft^3/s		
2	02098210	Haw River above Deep River near Haywood, NC	20.7	266	0.09	5.9	3.5 (G)	257	275	-2.9
3	02102000	Deep River at Moncure, NC	DA	86[1]	0.0	0.0	-- (P)	--	--	DA
F6		Hydropower power generation	NA	Power generation offline all day.						
4	0210204915	Deep River above mouth near Moncure, NC	DA	90.1	0.04	10	8.9 (P)	82	98	DA
5	02102050	Cape Fear River above powerplant intake near Moncure, NC	20.3	355	0.11	23	10.4 (P)	318	392	33.5
F7		Regional power utility	NA	Withdrawal: 213.82 ft^3/s				Return: 178.08 ft^3/s		
6	02102082	Cape Fear River below powerplant intake near Moncure, NC	19.4	217	0.05	10.5	7.9 (F)	197	231	-39.7
7	02102090	Cape Fear River above railroad bridge near Rosser, NC	18.4	211	0.05	10.9	8.3 (P)	194	228	-1.4
8	02102094	Cape Fear powerplant discharge canal near Brickhaven, NC	DA	170	0.68	2.9	2.7 (G)	165	175	DA
F8		Industrial	NA	Withdrawal: 0 ft^3/s				Return: No return discharge		
9	0210209450	Cape Fear powerplant discharge canal above juncture at Brickhaven, NC	DA	179	0.88	3.8	3.4 (G)	173	185	DA
10	0210209475	Cape Fear powerplant discharge canal below dam at Brickhaven, NC	DA	--	--	--	--	--	--	DA
11	0210209625	Unnamed stream between canal and Cape Fear River at Brickhaven, NC	DA	77	0.89	2.2	4.6 (G)	73	81	DA
12	0210209650	Cape Fear powerplant discharge canal below juncture at Brickhaven, NC	DA	98.3	0.73	1	1.7 (G)	97	100	DA
13	02102159	Lick Creek near Rosser, NC	DA	0.0	--	--	--	--	--	DA

Table 8. Results of synoptic discharge measurements made on September 9, 2009, on the Haw and Cape Fear Rivers between Jordan Lake Dam and Lillington, North Carolina.—Continued

[USGS, U.S. Geological Survey; main stem refers to the Haw and Cape Fear Rivers; ft³/s, cubic foot per second; change, in percent, computed as difference in measured discharge from next upstream measured discharge on main stem; DA, doesn't apply; NA, not available; --, not measured; measurement rating determined from base uncertainty at 95-percent confidence interval: (G), good; (F), fair; (P), poor (http://nc.water.usgs.gov/usgs/info/qaplan/surface.html)]

Site index number (figs. 1, 6, 7)	USGS station number	Station name	Main stem river miles upstream from Lillington streamgage (site 28)	Measured discharge, in ft³/s	Average velocity, in ft/s	Standard deviation, in ft³/s	95-percent confidence interval of measurement uncertainty, in percent (measurement rating)	Lowest discharge in 95-percent confidence interval, in ft³/s	Highest discharge in 95-percent confidence interval, in ft³/s	Change, in percent
14	0210215960	Cape Fear River above NC Highway 42 near Corinth, NC	16.7	--	--	--	--	--	--	DA
F9		Municipal	NA	Withdrawal: 11.28 ft³/s				Return: 0.61 ft³/s [2]		
15	0210215995	Cape Fear powerplant discharge canal at NC Highway 42 near Corinth, NC	DA	104	0.47	4	6.1 (F)	98	110	DA
16	02102176	Cape Fear powerplant discharge canal above mouth near Corinth, NC	DA	102	0.34	3.3	5.2 (F)	97	107	DA
17	02102177	Cape Fear River upstream from Buckhorn Dam near Corinth, NC	14.6	250	0.04	108	34.6 (P)	164	336	DA
Sites at or below Buckhorn Dam										
18	02102178	Cape Fear River at Buckhorn Dam near Corinth, NC	14.3	--	--	--	--	--	--	--
19	0210217810	Cape Fear River downstream from Buckhorn Dam near Corinth, NC	14	346	0.2	19.6	4.5 (G)	330	362	38.4
20	02102192	Buckhorn Creek near Corinth, NC	DA	0.0	--	--	--	--	--	DA
21	02102240	Cape Fear River adjacent to Bay Street near Cokesbury, NC	10	395	0.317	17.6	7.1 (F)	367	423	14.2
22	02102265	Cape Fear River at Raven Rock State Park group camp, NC	5.9	422	0.613	17.7	3.4 (G)	408	436	6.8
23	02102278	Cape Fear River above Hector Creek near Chalybeate, NC	4.2	424	0.423	29.8	5.6 (F)	400	448	0.5
24	02102280	Hector Creek near Chalybeate, NC	DA	4.67	0.147	--	--	--	--	DA
25	02102283	Cape Fear R adjacent to Bradley Road near Lillington, NC	3.4	404	0.26	17.6	7.0 (F)	376	432	-4.7
26	02102289	Cape Fear River above Neills Creek near Lillington, NC	1.4	358	0.297	14.7	3.3 (G)	346	370	-11.4
27	02102480	Neills Creek near Lillington, NC	DA	13.2	0.165	--	--	--	--	DA
F10		Municipal	NA	Withdrawal: 18.91 ft³/s				Return: Outside of study area		
28	02102500	Cape Fear River at Lillington, NC	0.0	403	0.3	41.6	8.3 (P)	370	436	12.6

[1] Computed daily discharge at this site for indicated date.

[2] Reported return discharge associated with water-treatment plant operations, not the wastewater-treatment operations that discharge to a tributary to the Deep River upstream from the study area.

Table 9. Results of synoptic discharge measurements made on October 1, 2009, on the Haw and Cape Fear Rivers between B. Everett Jordan Lake and Lillington, North Carolina.

[USGS, U.S. Geological Survey; main stem refers to the Haw and Cape Fear Rivers; ft³/s, cubic foot per second; change, in percent, computed as difference in measured discharge from next upstream measured discharge on main stem; DA, doesn't apply; NA, not available; --, not measured; measurement rating determined from base uncertainty at 95-percent confidence interval: (P, poor; (F, fair; (G, good (http://nc.water.usgs.gov/usgs/info/qaplan/surface.html)]

Site index number (figs. 1,6,7)	USGS station number	Station name	Main stem river miles upstream from Lillington streamgage (site 28)	Measured discharge, in ft³/s	Average velocity, in ft/s	Standard deviation, in ft³/s	95-percent confidence interval of measurement uncertainty, in percent	(measurement rating)	Lowest discharge in 95-percent confidence interval, in ft³/s	Highest discharge in 95-percent confidence interval, in ft³/s	Change, in percent
		Sites above Buckhorn Dam									
1	02098198	Haw R below B. Everett Jordan Dam near Moncure, NC	24.3	214	0.23	12	8.9	(P)	196	234	DA
F1		Industrial	NA	Withdrawal: 0.28 ft³/s					Return: 0.05 ft³/s		
F2		Industrial	NA	Withdrawal: 0 ft³/s					Return: 0 ft³/s		
F3		Industrial	NA	Withdrawal: 0.21 ft³/s					Return: 0.11 ft³/s		
F4		Industrial	NA	Withdrawal: 0.63 ft³/s					Return: 0.01 ft³/s		
2	02098210	Haw River above Deep River near Haywood, NC	20.7	204	0.08	17.9	7.0	(F)	190	218	-5.1
3	02102000	Deep River at Moncure, NC	DA	82 [1]	--	--	--		--	--	DA
F6		Hydropower power generation	NA	Power generation offline all day.							
4	0210204915	Deep River above mouth near Moncure, NC	DA	68.9	0.033	22.5	26.1	(P)	51	87	DA
5	02102050	Cape Fear River above powerplant intake near Moncure, NC	20.3	317	0.1	17.9	9.0	(P)	288	346	55.4
F7		Regional power utility	NA	Withdrawal: 395.77 ft³/s					Return: 316.87 ft³/s		
6	02102082	Cape Fear River below powerplant intake near Moncure, NC	19.4	46.2	0.01	20.5	35.5	(P)	30	63	-85.4
7	02102090	Cape Fear River above railroad bridge near Rosser, NC	18.4	40.8	0.01	30.9	60.6	(P)	16	66	-11.7
8	02102094	Cape Fear powerplant discharge canal near Brickhaven, NC	DA	309	0.95	7.7	4.0	(G)	297	321	DA
F8		Industrial	NA	Withdrawal: 0 ft³/s					Return: No return discharge		
9	0210209450	Cape Fear powerplant discharge canal above juncture at Brickhaven, NC	DA	308	1.25	5.5	2.9	(G)	299	317	DA
10	0210209475	Cape Fear powerplant discharge canal below dam at Brickhaven, NC	DA	--	--	--	--		--	--	DA
11	0210209625	Unnamed stream between canal and Cape Fear River at Brickhaven, NC	DA	159	1.51	8.2	4.1	(G)	152.	166	DA
12	0210209650	Cape Fear powerplant discharge canal below juncture at Brickhaven, NC	DA	154	0.84	5.6	5.8	(F)	145	163	DA
13	02102159	Lick Creek near Rosser, NC	DA	1.66	0.057	--	--		--	--	DA

Table 9. Results of synoptic discharge measurements made on October 1, 2009, on the Haw and Cape Fear Rivers between Jordan Lake Dam and Lillington, North Carolina.—Continued

[USGS, U.S. Geological Survey; main stem refers to the Haw and Cape Fear Rivers; ft³/s, cubic foot per second; ft/s, foot per second; change, in percent, computed as difference in measured discharge from next upstream measured discharge on main stem; DA, doesn't apply; NA, not available; --, not measured; measurement rating determined from base uncertainty at 95-percent confidence interval: (P), poor; (F), fair; (G), good (http://nc.water.usgs.gov/usgs/info/qaplan/surface.html)]

Site index number (figs. 1, 6, 7)	USGS station number	Station name	Main stem river miles upstream from Lillington streamgage (site 28)	Measured discharge, in ft³/s	Average velocity, in ft/s	Standard deviation, in ft/s	95-percent confidence interval of measurement uncertainty, in percent (measurement rating)	Lowest discharge in 95-percent confidence interval, in ft³/s	Highest discharge in 95-percent confidence interval, in ft³/s	Change, in percent
14	0210215960	Cape Fear River above NC Highway 42 near Corinth, NC	16.7	76	0.014	49.5	52.1 (P)	36	116	86.3
F9		Municipal	NA	Withdrawal: 10.85 ft³/s				Return: 0.09 ft³/s[1]		
15	0210215995	Cape Fear powerplant discharge canal at NC Highway 42 near Corinth, NC	DA	152	0.611	3.6	3.8 (G)	146	158	DA
16	02102176	Cape Fear powerplant discharge canal above mouth near Corinth, NC	DA	149	0.51	3.3	3.6 (G)	144	154	DA
17	02102177	Cape Fear River upstream from Buckhorn Dam near Corinth, NC	14.6	--	--	--	--	--	--	--
		Sites at or below Buckhorn Dam								
18	02102178	Cape Fear River at Buckhorn Dam near Corinth, NC	14.3							
19	0210217810	Cape Fear River downstream from Buckhorn Dam near Corinth, NC	14	285	0.18	71.6	26.4 (P)	210	360	DA
20	02102192	Buckhorn Creek near Corinth, NC	DA	0.8	0.03	--	--	--	--	DA
21	02102240	Cape Fear River adjacent to Bay Street near Cokesbury, NC	10	309	0.263	21	7.1 (F)	287	331	8.4
22	02102265	Cape Fear River at Raven Rock State Park group camp. NC	5.9	302	0.513	19.8	10.5 (P)	270	334	-2.3
23	02102278	Cape Fear River above Hector Creek near Chalybeate, NC	4.2	311	0.408	6.2	3.2 (G)	301	321	3.0
24	02102280	Hector Creek near Chalybeate, NC	DA	4.09	0.146	--	--	--	--	DA
25	02102283	Cape Fear R adjacent to Bradley Road near Lillington, NC	3.4	325	0.204	10.1	5.0 (G)	309	341	4.5
26	02102289	Cape Fear River above Neills Creek near Lillington, NC	1.4	318	0.262	10.3	5.2 (F)	302	334	-2.2
27	02102480	Neills Creek near Lillington, NC	DA	6.05	0.082	--	--	--	--	DA
F10		Municipal	NA	Withdrawal: 18.6 ft³/s				Return: Outside of study area		
28	02102500	Cape Fear River at Lillington, NC	0.0	327	0.24	11.4	5.6 (F)	309	345	2.8

[1] Computed daily discharge at this site for indicated date.

[2] Reported return discharge associated with water-treatment plant operations, not the wastewater-treatment operations that discharge to a tributary to the Deep River upstream from the study area.

During each series of synoptic measurements, discharge measurements also were made on selected tributaries where the contributing drainage area exceeded 5 percent of the drainage area of the main stem just upstream from the confluence. The selected tributaries include the Deep River, Lick Creek, Buckhorn Creek (fig. 6, sites 4, 13, 20, respectively), Hector Creek and Neills Creek (fig. 7, sites 24, 27, respectively). Measured discharges and average velocities are listed in tables 4–9. Not including the Deep River, measured discharge values at the four remaining tributaries were less than 5 percent and did not provide substantial flow contributions to the main stem, particularly during base-flow conditions.

In addition to the reduced flows downstream from the power-plant intake, average velocities from the measurements varied along the main stem (fig. 20; tables 4–9). Between the power-plant intake and Buckhorn Dam, the average velocities determined from the measurements were lower than along any other part of the main stem. For the synoptic measurements conducted during June through October, the average velocities in this reach were less than 0.1 foot per second (ft/s). In addition to the increased channel width between the intake point and Buckhorn Dam, the amount of storage in the channel upstream from Buckhorn Dam results in very slow water travel along this reach during base-flow conditions. The average velocities in the diversion canal adjacent to the main stem commonly ranged from 0.5 to about 1.5 ft/s at the canal-measurement sites (tables 4–9). Average velocities in the reach downstream from Buckhorn Dam increased from about 0.2 or 0.3 ft/s to a range of about 0.5 to almost 0.7 ft/s at the measurement sites at Raven Rock (fig. 7, site 22) and above Hector Creek (site 23), then decreased to a range of 0.2 to 0.4 ft/s toward the Lillington streamgage (site 28). These variations in average velocity downstream from Buckhorn Dam were attributed to changes in channel widths on the main stem between Raven Rock State Park and the Lillington streamgage (fig. 20) as well as minor variations in the channel slope (fig. 2). While changes in velocities do not factor into a determination of flow loss, an understanding of the velocity patterns determined from the series of synoptic measurements emphasizes recognition of the complex flow patterns in the reach.

Streamgage Data Collection and Analyses

In the years leading up to this study, questions were raised about the accuracy of the discharge records at the USGS streamgages used in the determination of possible flow losses. Questions also were raised concerning the determination of flow releases for Jordan Lake Dam operated by the USACE. Therefore, a comparison of the stage-discharge ratings and the measured discharges used to compute discharge records was completed for Deep River at Moncure (fig. 1, site 3) and Cape Fear River at Lillington (fig. 1, site 28). In addition to the rating analyses for these two sites, velocity data at the streamgage on the Haw River below B. Everett Jordan Dam (site 1) downstream from Jordan Lake and stage data at Cape Fear River at Buckhorn Dam near Corinth (site 18)

were collected. Analyses completed on the ratings at the two streamgages (sites 3 and 28) did not indicate the presence of bias in the discharge computations that would result in the occurrence of false flow losses. Comparisons of discharge measurements collected on the Haw River below Jordan Lake Dam provided general support of the current USACE discharge-computation tables used for reporting flow releases from the dam.

Analyses of Ratings and Measured Discharges

Analyses of discharge measurements and ratings for the two streamgages at Deep River at Moncure (site 3) and Cape Fear River at Lillington (site 28) were completed as part of this investigation to verify the accuracy of the computed discharges. USGS personnel responsible for the operation of the streamgages were routinely called upon to make special visits to the sites to verify their accuracy and functionality, as was the case for the Lillington streamgage (site 28) during the years leading up to this study. The analyses were completed to determine if a systematic bias existed in the discharge records, possibly giving a false computation of flow loss between upstream and downstream ends of the study reach. For the streamgage at Cape Fear River at Lillington (site 28) at the outlet of the study reach, such a bias would be indicated by consistently positive percentage differences over time (following any rating shifts), meaning the base rating discharges would be consistently lower than the measured discharges. Such a pattern could potentially result in an apparent flow loss due to differences between the measured discharge and the discharge computed using the stage-discharge rating. However, for the streamgage at Deep River at Moncure (one of the inputs, site 3), a bias would be indicated by shifted percentage differences that are consistently negative, meaning the base rating discharges would be consistently higher than the measured discharges obtained during the streamgage inspections.

Assessment of stage-discharge ratings focused on discharge measurements and ratings after February 1982, the month during which lake levels at Jordan Lake initially reached normal pool elevation. Two ratings were used during the 1982–2011 water years at Deep River at Moncure (site 3). A total of 99 discharge measurements were made at this site during the same period and ranged in magnitude from 27.9 to 21,500 ft^3/s (fig. 21A). The measurements made during this period were rated as good (40 percent), fair (44 percent), or poor (16 percent). Among these measurements, 66 were below 500 ft^3/s; the distribution of these measurement ratings as good, fair, or poor was 25, 55, and 20 percent, respectively.

Percentage differences at Deep River at Moncure (site 3) during the 1982–2011 water years ranged from about –7.5 percent to nearly +19 percent with values within plus or minus (+/–) 5 percent for 85 measurements (fig. 21B). For the measurements followed by a temporary rating shift (indicated by X symbol in fig. 21B), the percentage differences indicated are the shifted percentage differences. The measurement with a percentage difference of almost 19 percent was rated as

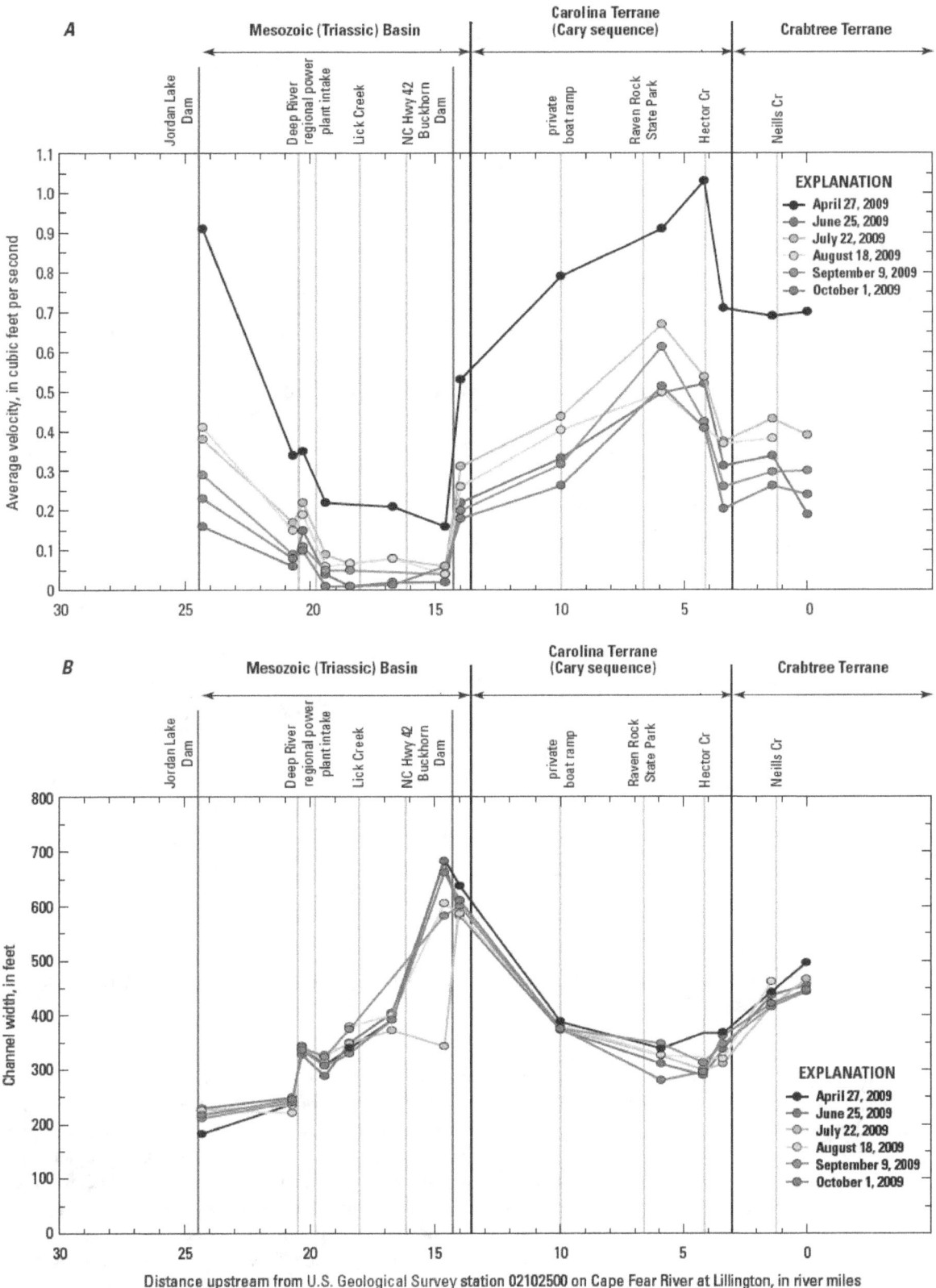

Figure 20. (*A*) Average velocities and (*B*) channel widths determined from the series of synoptic measurements made during 2009 on the Haw and Cape Fear Rivers between Jordan Lake Dam and Lillington, North Carolina.

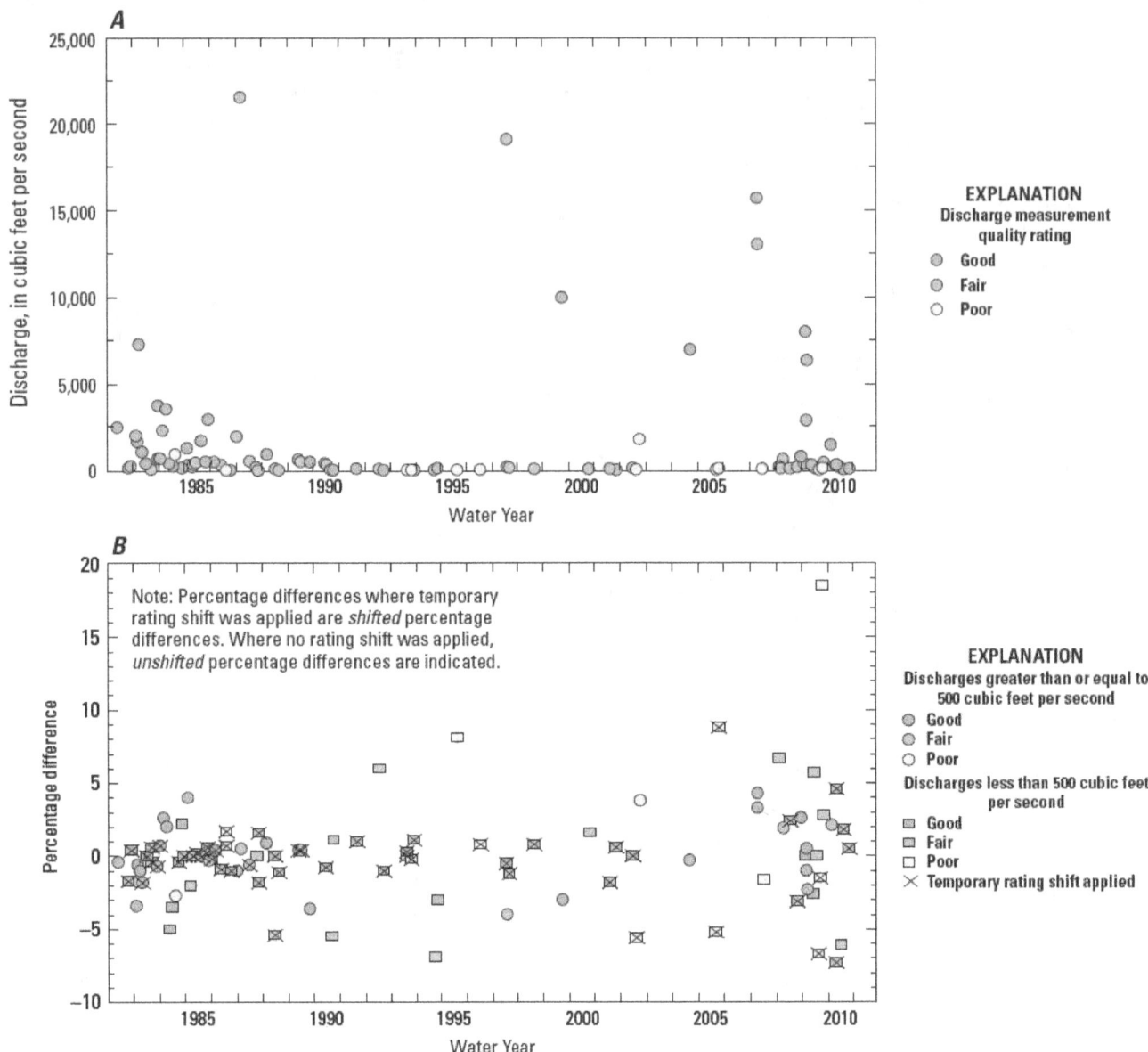

Figure 21. Quality ratings of (*A*) discharge measurements and (*B*) percentage differences from shifted, where applied, base rating at Deep River at Moncure (site 3), North Carolina, during the 1982–2011 water years.

poor because of equipment malfunction, and the measurement was not used to apply a rating shift. A total of 53 rating shifts were applied during the period, and 17 shifts were positive and 36 were negative (fig. 21*B*). Among the 66 measurements less than 500 ft³/s, rating shifts were applied after 45 measurements were completed, and 14 shifts were positive and 31 were negative. Whether shifted or unshifted, percentage differences at the Moncure streamgage do not indicate a particular time during the 1982–2011 water years when the values were either consistently positive or negative (fig. 21*B*). The majority of the percentage differences and corresponding shifts were negative, which would support the hypothesis that possible false computations of flow loss occur due to the discharge records at the Moncure streamgage. However, when examining the temporal trend in the

percentage differences for measurements less than 500 ft³/s, there was no indication of a long-term temporal bias in which the values were negative (fig. 21*B*). The absence of negative bias indicates that flow losses in the study reach are not attributable to the computation of discharge records at the Moncure streamgage.

Four ratings have been used for the Cape Fear River at Lillington streamgage (site 28) since December 1976. A total of 103 discharge measurements made at this streamgage during the 1982–2011 water years range from 133 to 28,800 ft³/s (fig. 22*A*). The measurements made during this period were rated as excellent (1 percent), good (47 percent), fair (42 percent), or poor (11 percent). Among these measurements, 51 were measured at discharges of 1,000 ft³/s or less. Among these 51 discharge measurements, the distribution of

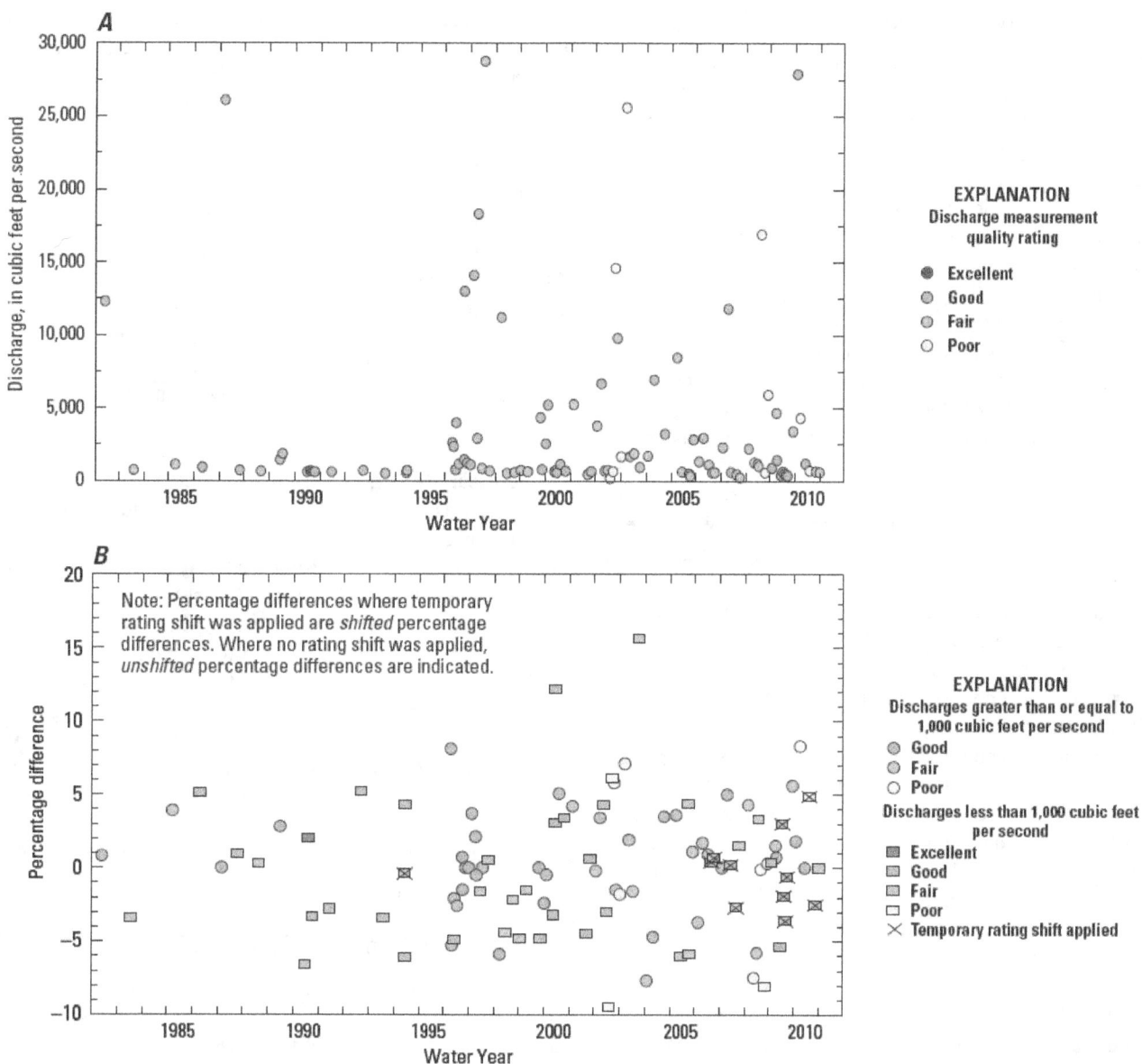

Figure 22. Quality ratings of (*A*) discharge measurements and (*B*) percentage differences from shifted, where applied, base rating at Cape Fear River at Lillington (site 28), North Carolina, during the 1982–2011 water years.

measurement ratings as excellent, good, fair, or poor was 2, 43, 47, and 8 percent, respectively.

Percentage differences for the 103 measurements at the Cape Fear River at Lillington streamgage (site 28) during the 1982–2011 water years ranged from about –10 percent to about +16 percent, with values within +/– 5 percent for 79 of the 103 measurements (fig. 22*B*). A total of 24 rating shifts were applied, and 23 shifts were positive and 1 was negative (fig. 22*B*). Among the 51 measurements less than 1,000 ft³/s, rating shifts were applied after 11 measurements were completed, and 10 shifts were positive and 1 was negative. The percentage differences, whether shifted or unshifted (shifted values are indicated by X symbol in fig. 22*B*), likewise do not indicate a particular time during the 1982–2011 water years when the values were either

consistently positive or negative (fig. 22*B*). When examining just the 51 measurements less than 1,000 ft³/s, there is likewise no indication of bias in the percentage differences during this period. The majority of shifts were positive, which lends support to the hypothesis that possible false computations of flow loss occur due to the discharge records at the Lillington streamgage. However, relatively few shifts were applied at this site; all but one of the shifts occurred during and following the 2001 water year (fig. 22*B*). Where shifts were not applied to the rating, the current ratings in effect at the time of measurement were regarded as appropriate for the continual computation of discharge records. The absence of shifts further indicates that no long-term temporal bias has occurred in the ratings at the Lillington streamgage that could result in false computation of flow losses.

Streamgage at Haw River below B. Everett Jordan Dam

Questions concerning suspected flow losses in the years prior to this investigation indicated some concerns about the accuracy of reported flow releases from Jordan Lake Dam. To assess the accuracy of flow releases, a streamgage on the Haw River below B. Everett Jordan Dam near Moncure (site 1), operated since 1992 for stage-only records, was converted to a discharge-record site in November 2008. Streamflow records for this site were published from October 1965 to October 1992. Prior to October 1978, the site was operated 0.3 mi downstream from the dam with records published under the name of Haw River near Haywood (USGS station 02098200).

The effects of storage behind Buckhorn Dam extend upstream to Jordan Lake Dam. The result of this storage creates a backwater effect that prevents a normal stage-discharge relation from being established. High flows traveling down the Deep River also create a backwater effect on Haw River flows immediately downstream from Jordan Lake Dam. When the Haw River streamgage was operated previously as a discharge station, the site was operated as a "slope station" because an auxiliary gage was located downstream from the primary or base streamgage. Stage records collected at both locations allowed the determination of a slope between the two points, a variable used in the computation of discharge records.

The streamgage operated at this location during the study was an index-velocity site, whereby continuous velocity records are collected in one or more volume samples in the channel and used to establish a relation with the average velocities determined by individual discharge measurements. A second relation is developed between water stage and cross-sectional area of the channel. Discharge is then computed by multiplying the average velocity from the first relation by the area determined from the second relation.

Channel velocities are collected by using acoustical transducers that emit sound waves into the channel; returning sound waves from moving particles in the flow are used to compute an index velocity based on the Doppler principle. Flow releases from the dam often resulted in entrained air bubbles in the water column at the cross section where the velocity transducers were located. When this phenomenon occurred, the velocity data displayed such a wide range in variability that questions arose concerning the ability of the instrumentation to obtain accurate velocity data needed to compute the discharge. Following a review (during the latter part of the investigation) of the index-velocity data collected at this streamgage, records of daily discharge computed for this site were deemed unreliable. Nearing the end of the investigation, plans were underway to relocate the streamgage approximately 1.25 mi downstream at the U.S. Highway 1 bridge where reconnaissance discharge measurements indicated more suitable conditions for index-velocity streamgage operations were likely.

Following the review, a focus was placed on comparison of individual measured discharges and the reported hourly flow releases from Jordan Lake Dam. A total of 34 discharge measurements were collected during the study at the streamgage downstream from Jordan Lake Dam (site 1). Measured discharges ranged from 94.7 to 9,970 ft^3/s (table 10; fig. 23A), and 25 of these measurements were less than 2,000 ft^3/s. Comparison of the USGS measured discharges and USACE-supplied flow releases indicated that the measured discharges ranged from 75 to about 140 percent of the concurrent hourly flow releases among the 34 measurements, and the values were within +/– 10 percent for 28 measurements. Among the 6 measurements with differences greater than 10 percent, explanation for the differences lies in changes in hourly flow releases close to the time of measurement.

Flow releases from Jordan Lake Dam were made through service gates located in the intake tower immediately upstream from the dam. Tables of predetermined discharges based on lake level and gate openings are used to determine the opening necessary for a given discharge. Adjustments to the gate openings are completed using mechanical adjustment and visual inspection of a scale linked to each opening. When attempting to set the flow for a relatively small release amount (for example, during drought conditions), adjustments to the gates may be a matter of 1 to 2 in., making verification of the desired opening more challenging for personnel. According to the USACE, at a lake level of 216 ft above NGVD 29 (and at lower levels), each 1-in. gate opening equates to an approximate difference of 30 ft^3/s in discharge magnitude (Ashley Hatchell, U.S. Army Corps of Engineers, written commun., March 2, 2010). The occurrence of 28 of 34 measurements being within +/- 10 percent of the reported hourly flow releases provides general support of the current discharge computation tables used for reporting Jordan Lake Dam flow releases.

Stage-Only Streamgage on Cape Fear River at Buckhorn Dam

An absence of stage data on the Cape Fear River at Buckhorn Dam raised questions about whether flow losses could be attributed to fluctuations in the water surface at this structure. In other words, the question to be answered is whether the water surface declines to levels below the spillway crest, thereby resulting in flow losses measured at the Lillington streamgage (site 28).

During the study, a stage-only streamgage was operated on the Cape Fear River at Buckhorn Dam near Corinth (site 18) to collect continuous stage data. No records of discharge were collected at this site. Except for one week during January 2009, the station was operated from November 24, 2008, to June 25, 2010, when the gage and recording equipment were lost due to theft. A temporary stage gage was installed to continue data collection during the latter stages of the investigation. Data from the temporary gage provided additional data through July 31, 2010. Stage records were converted into elevations following a survey that determined the station datum for the site, which was referenced to the elevation of a temporary benchmark established using Real-Time Kinematic (RTK) GPS equipment. The elevation of the station datum was determined to be 156.49 ft above NAVD 88

Table 10. U.S. Geological Survey discharge measurements made at Haw River below B. Everett Jordan Lake near Moncure, North Carolina, and reported hourly flow releases from Jordan Lake Dam, December 2008 through December 2010.

[ft³/s, cubic feet per second; measurement rating determined from base uncertainty at 95-percent confidence interval: (G), good; (F) fair; (P), poor (http://nc.water.usgs.gov/usgs/info/qaplan/surface.html); hrs., hours; NR, none recorded]

Date	Time of measure-ment	Measured discharge, in ft³/s	Stage, in feet	Measurement rating	Reported hourly flow release, in ft³/s	Time	Comments
12/3/2008	1058	1,710	4.66	G	1,573	1100	NR
12/15/2008	1138	1,690	4.54	G	1,407	1100	Substantial adjustments in flow release between 1000 and 1500 hrs (from 303 to 4,592 ft³/s).
12/15/2008	1212	3,300	5.88	G	3,024	1300	Substantial adjustments in flow release between 1000 and 1500 hrs (from 303 to 4,592 ft³/s).
12/15/2008	1455	5,030	7.16	G	4,592	1500	Substantial adjustments in flow release between 1000 and 1500 hrs (from 303 to 4,592 ft³/s).
3/4/2009	1109	4,220	8.60	G	4,242	1200	Measurement made during increase in stage from 8.23 ft at 1100 hrs to 8.64 ft at 1115 hrs to 8.74 ft at 1130 hrs.
3/5/2009	1137	6,200	9.00	G	6,080	1100	NR
3/17/2009	1057	1,450	6.18	G	1,550	1100	Substantial adjustment in flow release between 1300 and 1400 (from 608 to 2,512 ft³/s).
4/22/2009	1010	838	3.90	G	788	1000	NR
4/27/2009	1233	859	3.74	G	787	1200	NR
6/25/2009	1052	164	3.19	F	130	1100	NR
7/22/2009	0750	430	3.34	G	420	0800	NR
8/18/2009	1015	443	3.28	G	417	1000	NR
9/9/2009	0926	274	3.18	P	288	0900	NR
9/30/2009	1033	94.7	3.03	P	127	1100	Flow releases from Jordan Lake dam adjusted between 1000 and 1400 hrs.
9/30/2009	1122	169	3.06	F	127	1100	Flow releases from Jordan Lake dam adjusted between 1000 and 1400 hrs.
9/30/2009	1142	167	3.05	F	127	1100	Flow releases from Jordan Lake dam adjusted between 1000 and 1400 hrs.
9/30/2009	1224	270	3.08	F	192	1200	Flow releases from Jordan Lake dam adjusted between 1000 and 1400 hrs.
9/30/2009	1244	273	3.06	G	192	1200	Flow releases from Jordan Lake dam adjusted between 1000 and 1400 hrs.
10/1/2009	0850	214	3.10	F	223	0800	NR
11/3/2009	1018	193	3.14	F	193	1000	NR
11/6/2009	0858	134	3.15	G	128	0900	NR
11/6/2009	0948	124	3.13	G	128	0900	NR
11/17/2009	1105	7,570/s	9.82	G	7,404	1100	NR
11/17/2009	1124	7,470	9.82	G	7,404	1100	NR
11/19/2009	0944	9,970	11.48	G	9,950	0900	NR
11/19/2009	1000	9,830	11.46	G	9,946	1000	NR
1/13/2010	1107	419	3.37	F	393	1100	NR
2/23/2010	0950	2,830	5.86	G	2,958	1000	NR
2/26/2010	1110	124	3.80	G	131	1100	NR
4/22/2010	0934	414	3.37	G	425	0900	NR
6/8/2010	1350	693	3.56	G	658	1300	NR
8/11/2010	1020	487	3.31	G	522	1000	NR
8/11/2010	1059	501	3.31	G	522	1100	NR
12/8/2010	1804	380	3.34	G	393	1800	NR

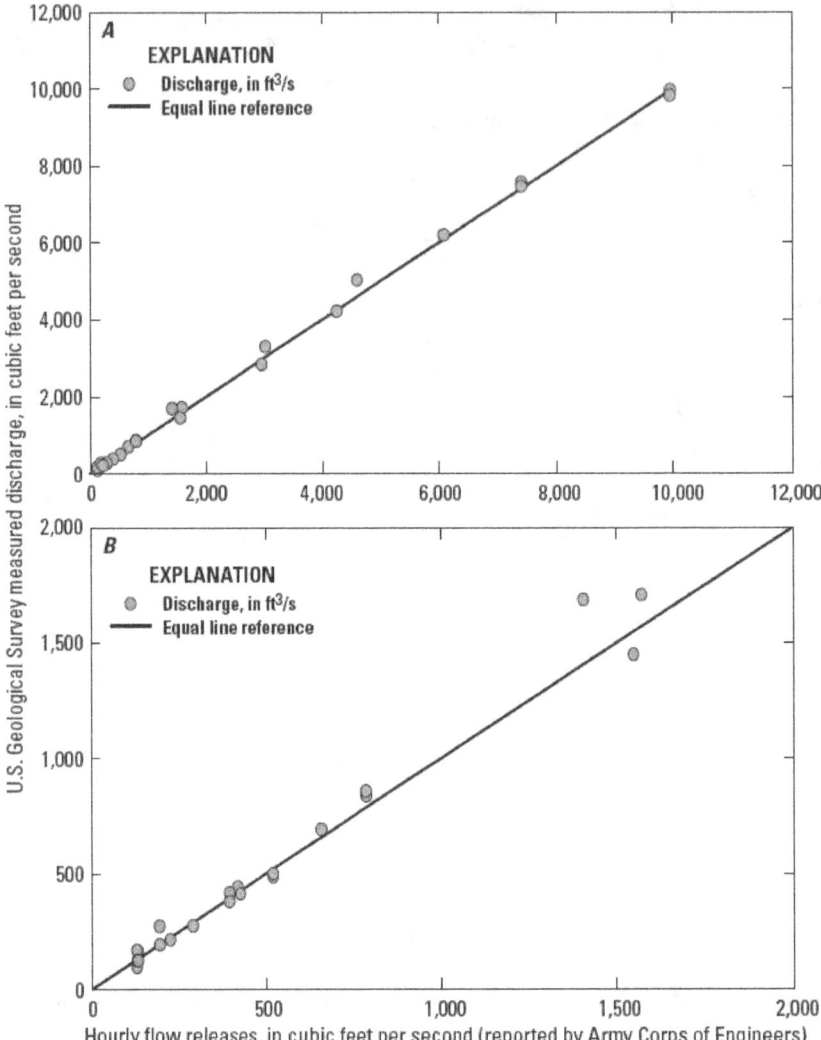

Figure 23.   (*A*) Full range of measured discharges at Haw River below B. Everett Jordan Lake near Moncure (site 1), North Carolina, and (*B*) measured discharges less than 2,000 cubic feet per second.

with an accuracy of 0.15 ft. The left end of the dam crest was surveyed at an elevation of 157.25 ft.

The crest length of Buckhorn Dam is almost 1,700 ft with a structural height of 25 ft (North Carolina Department of Environment and Natural Resources, 2010). Aerial photographs (from Google Earth) indicate the crest of the dam between the edges of the river is approximately 1,100 ft in length.

Stage data collected during the study and converted to a continuous record of elevation ranged from 157.32 ft on October 9 and 12, 2009, to 161.42 ft on February 6, 2010 (fig. 24). Throughout the study period, flow over the dam was observed along its length and is confirmed by the minimum recorded elevation being at a slightly higher elevation than that surveyed for the left end (facing downstream) of the dam crest. During the nearly 20-month period of record for this station, there were 14 days when flow loss was computed using the methods described in previous sections. Among these 14 days, the stage

ranged from 0.14 to 1.89 ft above the left end of the dam crest. This range indicates that flow loss in the study reach is not attributed to a decrease in stage below the dam crest.

Water Use

Developing an understanding of the flows in the study area required an assessment of available water-use records of diversions that affect the discharges in the Haw and Cape Fear Rivers. The diversion records for the study area, and particularly along the main stem, raised questions in the year prior to the study about whether flow losses could be attributed to water use. Using the median flow loss of 37 ft^3/s (table 3; flow comparison 3, daily discharge) as a point of reference, water-use analyses completed during the investigation indicated that total diversions were equivalent to 69 and 104 percent of the median flow loss during the winter and summer periods, respectively.

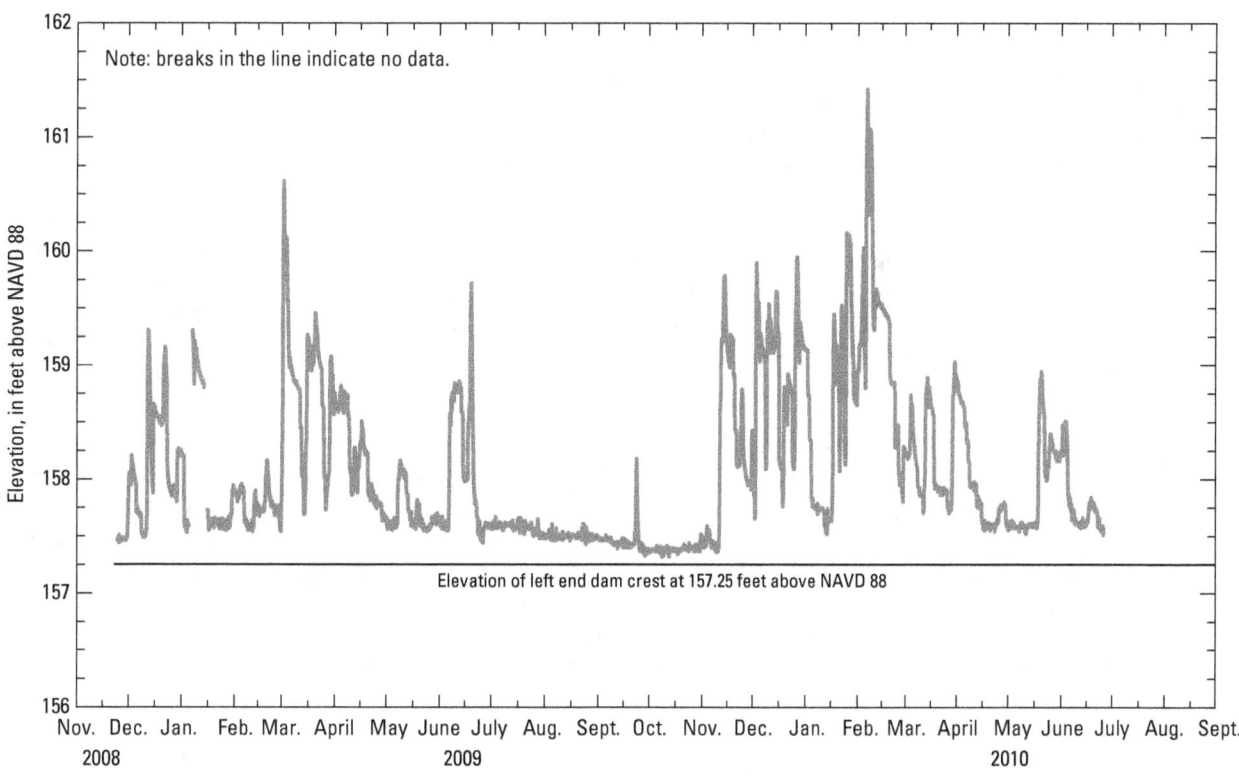

Figure 24. Instantaneous elevation of water surface on the Cape Fear River at Buckhorn Dam near Corinth (site 18), North Carolina, November 2008 through July 2010.

Near the beginning of the study, an initial assessment of water users within or near the study area was completed (Don Rayno, North Carolina Division of Water Resource, oral commun., May 28, 2009). Using information from this assessment, water-use information and (or) data were obtained for five industrial facilities—a regional power utility, two municipalities, one small hydropower facility on the Deep River, and one quarry operation adjacent to the Deep River (table 11). Data indicate that the level of water consumption—the portion of water withdrawals not returned to the rivers by way of return point-source discharges—varies among these facilities.

The compilation of water-use data indicates that the largest users are the regional power producer, the small hydropower facility that operates in a run-of-river mode, and the two municipalities (table 11). Water use at the regional power plant and small hydropower facility is for power-generation purposes. Only part of the water through the regional power plant is consumed, and no water consumption occurs in the run-of-river operations at the hydropower facility. Water use at these two facilities is discussed in further detail in subsequent sections.

Most of the facilities do not directly measure the water withdrawals from the rivers. Rather, water use between an off-stream storage location and the facility is compiled and made available for reporting to regulatory agencies. Water use at both municipal intakes is measured between the storage pond and the water-treatment facilities. Expansion of the water-treatment plant at one of the municipal facilities was underway during

the investigation. By the end of the study, pump flowmeters were in position to directly measure future water withdrawals from the Cape Fear River. Likewise, water-use information available for four of the five industrial users was monitored between an onsite fire pond and (or) storage facility and the facility itself. Paper records of direct water withdrawals from the river are available for one industrial facility (fig. 6, site F4); however, major facility expansion during the investigation altered the daily operations such that water-use data reported for the synoptic series are not typical of normal operations. Among the privately owned facilities for which water-use data were sought during the investigation, only data on a daily scale were obtained in electronic format for facilities F1 and F7 for 2006–2010 (fig. 25A) and 2008–2010 (fig. 25B), respectively. Changes in daily water use at facility F1 that were attributed to the economic recession during the study period were noted with a decrease from 0.4–0.5 ft³/s to 0.3–0.4 ft³/s beginning in early 2009 (fig. 25A). Water use at the regional power plant (fig. 25B) is discussed further in a subsequent section.

A quarry operation (fig. 1, site F5) is located adjacent to the Deep River just downstream from the streamgage on the Deep River at Moncure (site 3). Information obtained from the environmental compliance representative for the quarry indicated that reported water withdrawals are from the quarry pit, with most of the withdrawn water attributed to the accumulation of precipitation runoff in the quarry pit (David Lee, Wake Stone Corporation, Geologist/Environmental Supervisor oral commun.,

Table 11. Summary of selected surface-water withdrawals and return point-source discharges in the Cape Fear study reach, North Carolina.

[ft³/s, cubic foot per second; Mgal/d, million gallons per day (1 Mgal/d is equivalent to 1.5472 ft³/s); NPDES, National Pollutant Discharge Elimination System; <, less than; ~, approximately; NA, not applicable; ND, no data; USGS, U.S. Geological Survey; WTP, water-treatment plant; WWTP, wastewater-treatment plant. Facility name and specifc diversion location not identified in accordance with federal security regulations. Reported withdrawals or water usage for facililties are based on electronic records (period indicated in Remarks) and (or) anecdotal information provided by facility]

Site index number (figs. 1, 6, 7)	Facility type	Source of water supply	Withdrawals or water usage, ft³/s (Mgal/d)	Destination of return point-source discharge	Return point-source discharge, ft³/s (Mgal/d)	Permitted NPDES discharge, ft³/s (Mgal/d)
F1	Industrial	Haw River	0.3 to 0.45 (0.2 to 0.3)	Haw River	0.08 to 0.15 (0.05 to 0.1)	0.37 (0.24)

Remarks: Facility manufactures polyester fiber used in tire cord market. Water usage and point-source discharge data are based on electronic records obtained for 2006-2010 calendar years (fig. 25A). Water usage since early 2009 has declined to about 0.2 Mgal/d due to economic downturn.

F2	Industrial	Haw River	< 0.1 (< 0.06)	Haw River	See Remarks	0.01 (0.006)

Remarks: Facility manufactures plywood used for wood products. Water is pumped from Haw River into fire pond with overflow to the Haw River; pump is manually turned on and off as needed. No records of water withdrawals or usage or pump run times are maintained. Pump capacity was stated to be about 300-400 gallons per minute. Return point-source discharage to Haw River from on-site waste pond is managed for company by way of contract and is not continuous, being used as needed. The company is advised by the contractor when to complete a return discharge.

F3	Industrial	Haw River	< 0.4 (< 0.25)	Haw River	< 0.15 (< 0.1)	0.15 (0.1)

Remarks: Facility manufactures resins used in the furniture industry. Water usage and point-source discharge data are based on information obtained during on-site visits. No records of water withdrawals or usage or pump run times are maintained. Two pumps are used to withdraw water (rated 300 gallons per minute, but more likely running at about 170 to 180 gallons/min due to wear and tear on pumps over time).

F4	Industrial	Haw River	~ 1.5 (~ 1.0)	Haw River	0.03 (0.02)	Not limited

Remarks: Facility manufactures particle board and medium density fiber board. Water usage and point-source discharge data are based on information obtained during on-site visits. Paper records are maintained in on-site files, but no electronic data were available for water withdrawals or usage. Average daily withdrawal in 2008 reported to N.C. Division of Water Resources ranged from 0.45 to 1.5 ft³/s (0.3 to 1.0 Mgal/d). Major facility expansions were underway during the study period. Following the expansions, daily withdrawals are projected to increase to about 3 ft³/s (2 Mgal/d), and the return point-source discharge likely will increase to about 0.4 ft³/s (0.25 Mgal/d). Part of treated water is spray discharged onto adjacent fields owned by company.

F5	Quarry	Water collected at bottom of quarry	~ 0.2 to ~ 0.45 (0.14 to 0.3)	Unnamed tributary to Deep River	0 to ~ 0.45 (0 to 0.3)	See Remarks

Remarks: Water usage and point-source discharge data are based on information for 2009 reported to N.C. Division of Water Resources. Return point-source discharge occurs under Stormwater General Permit with no specified discharge limit. According to quarry management, most of water withdrawn from quarry is rainwater runoff collected at bottom. Water withdrawn from the quarry is discharged to an unnamed tributary to the Deep River adjacent to the quarry or routed to a process-water reservoir for later use in stone washing and dust suppression operations.

F6	Hydropower power generation	Deep River	~ 250 to ~ 550 (~ 160 to ~ 355)	NA	NA	See Remarks

Remarks: Small hydropower operation is located about 1 mile downstream from USGS streamgage on Deep River near Moncure (site 3). General range in water use corresponds to discharges through turbine when in operation. Water is return discharged to the Deep River after passage through turbine. Water usage information listed for this facilty is estimated flow in 0.5-mile canal between dam and powerhouse. Water is return discharged to Deep River by way of flow through turbine. See text for further discussion concerning this facility.

F7	Regional power utility	Cape Fear River	~ 290 to ~ 400 (~ 190 to ~ 260)	Diversion canal adjacent to Cape Fear River	ND	Not limited

Remarks: Facility is largest water user in the study area. Withdrawal information is based on daily average of 15-minute interval intake volumes estimated using adjusted pump curves developed by regional power producer in late 2010 (fig. 25B). Pumping varies during the year according to power demands, highest in the summer and lowest during the winter. No long-term information was obtained on the return point-source discharge, which is located downstream from USGS streamgage on discharge canal adjacent to the Cape Fear River (site 8), because of the presence of this gaging station. See text for further discussion concerning this facility.

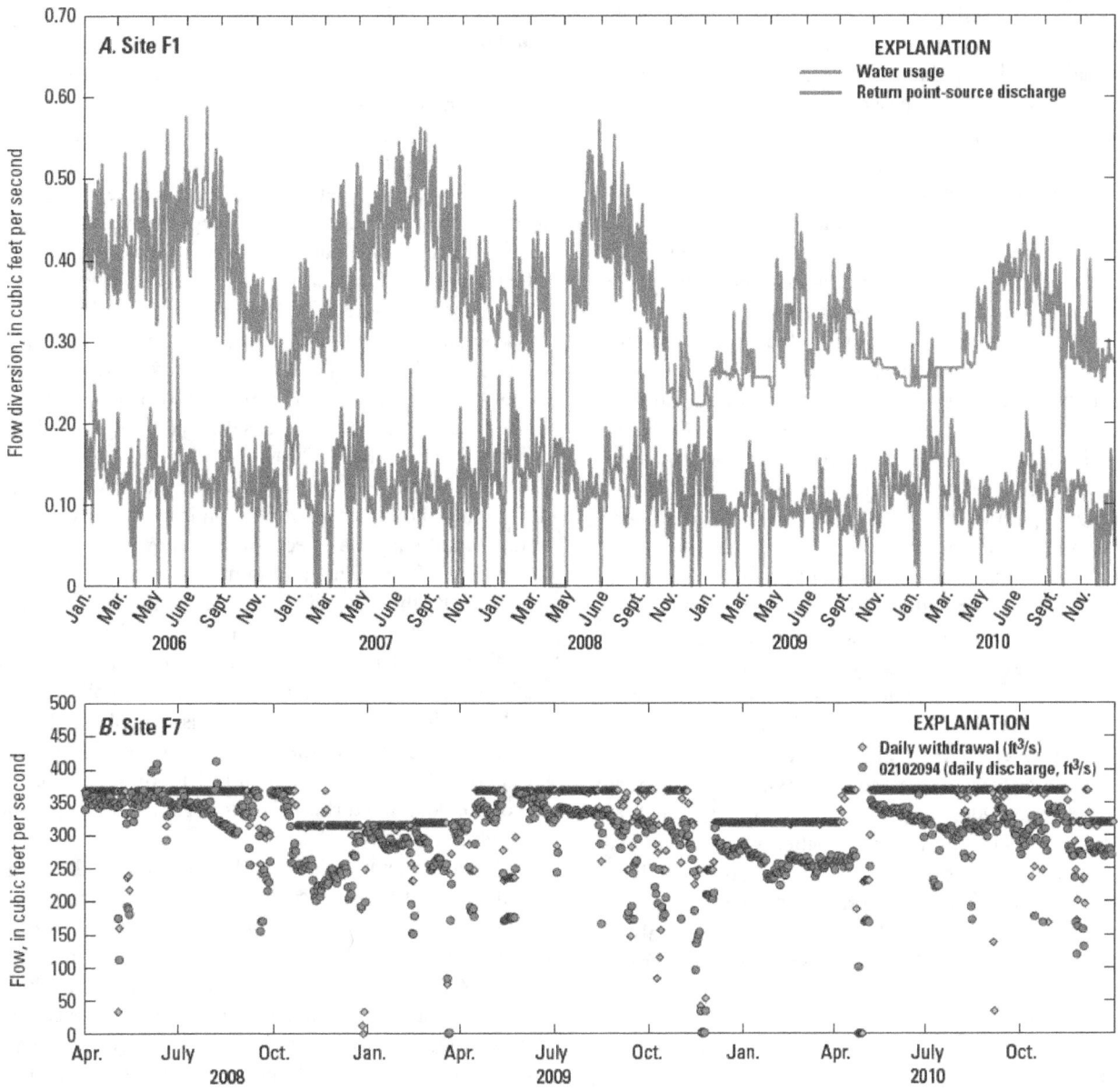

Figure 25. Flow diversions at (*A*) industrial facility (site F1) on the Haw River during 2006–2010, and (*B*) regional power plant (site F7) during 2008–2010, North Carolina.

May 17, 2010). The reported point-source discharge is to an unnamed tributary to the Deep River. No daily records of water withdrawal were obtained during the study; however, the 2009 annual water-use report filed with the North Carolina Division of Water Resources indicated that both average daily water withdrawals and return point sources were less than 0.3 Mgal/d (0.45 ft³/s; North Carolina Division of Water Resources, 2009).

Information concerning irrigation withdrawals in North Carolina has been limited historically, as State regulations exempt agricultural operations from reporting requirements for water withdrawals and usage. However, one exception is a 2007 report filed with the State agency for a farm operation near the Cape Fear River near Lillington (Don Rayno, North Carolina Division of Water Resources, oral commun., May

28, 2009). Average daily withdrawals during June through August were reported at about 1.2 Mgal/d (about 2 ft³/s; North Carolina Division of Water Resources, 2007). No observations of irrigation piping into the main stem were observed in field operations during the study period. Aerial photographs indicate limited occurrences of agricultural land cover immediately adjacent to the main stem.

The combined maximum daily water withdrawal or usage is about 3 ft³/s (about 2 Mgal/d) for the five industries and quarry operation. Similarly, the combined minimum return point-source discharge is less than 0.5 ft³/s (about 0.3 Mgal/d), implying the maximum water consumption among the six facilities would be about 2.5 ft³/s (about 1.6 Mgal/d). Even adjusting for the projected increases in water withdrawals and return discharges at facility F4 following major expansions during the study period, the maximum water consumption among the industrial users is less than 4.0 ft³/s (about 2.6 Mgal/d).

Water withdrawals at the two municipal water-supply intakes (sites F9, F10) represent 100 percent consumption in the study area because the return point-source discharges from corresponding wastewater-treatment plants are located outside of the study area. The return point-source discharge for facility F9 is on a tributary to the Deep River upstream from site 3 near Moncure (fig. 6), and the return point-source discharge for facility F10 is downstream from site 28 at Lillington (fig. 7). Monthly (facility F9) and daily (facility F10) water-use data in electronic format were available for these two intakes and provided some insight about variations in water use through the calendar year. As with most of the other industries examined during the investigation, water use is monitored between a storage reservoir and the water-treatment plants for both municipalities, although major expansion completed during the study period at one of the water-treatment plants (facility F10) included the installation of pump flowmeters for monitoring future water withdrawals directly from the river.

Since 2000, the amount of water use associated with the upstream municipal intake (facility F9) has ranged from 9 to 12 ft³/s (5.8 to 7.8 Mgal/d) during the winter and summer periods, respectively (fig. 26; table 11). Of note in the data for 1986–2010 calendar years provided for this municipal intake, average water use increased approximately 67 percent between 1986 and 2000, rising from 6 to 10 ft³/s (about 3.9 to 6.5 Mgal/d). Data indicate water-use patterns have been level since 2000 with a slight decline during 2005–2010, a reflection of changes in water use following recent drought periods.

Water use associated with the downstream municipal intake (facility F10) during 2006–2010 typically ranged from 14 to 24 ft³/s (9.0 to 15.5 Mgal/d) during the winter and summer periods, respectively (fig. 26; table 11). Following the recent expansions at the downstream municipal intake (facility F10), increased pumping at this facility alone ranged from 24 to about 35 ft³/s (15.5 to 22.6 Mgal/d) beginning in early 2010 (fig. 26).

Combining the range of withdrawals for just the two municipalities indicates that from 23 to 36 ft³/s (35.6 to 55.7 Mgal/d) is removed from the study area during the winter and summer periods, respectively. When adding the maximum water consumption of about 2.5 ft³/s (about 1.6 Mgal/d) for the five industries and quarry operation, the total water-use diversions can range from almost 25.5 to 38.5 ft³/s (39.5 to 59.6 Mgal/d) during the winter and summer periods, respectively. This range in water-use diversions is equivalent to 69 to 104 percent of the 37 ft³/s median flow loss (table 3, flow comparison 3, daily discharge).

Lockville Hydropower Station

The Lockville hydropower station (fig. 6, site F6) is on the Deep River about 1 mi downstream from the Moncure streamgage (site 3). A small low-head dam across the river, constructed in the mid-1800s, is 0.5 mi upstream from the

powerhouse (built in the early 1920s) and connected by a canal built between the two structures. Ownership of the dam, canal, and powerhouse has changed hands during previous decades and most recently in 2003. No hydropower generation occurred from 2003 to early 2008 while equipment repairs and upgrades were completed by the current (2011) owner.

Hydropower generation at the station, regulated for operation by the Federal Energy Regulatory Commission (FERC), occurs on a continual basis except during instances when insufficient streamflow prevents operation of the turbines or when operations have been stopped for the repair or protection of equipment. The powerhouse contains two turbines capable of generating 750 kilowatts (kW) of power, but only one is being used currently (2011) to generate power (William Brooks, owner, Lockville hydropower station, oral commun., November 9, 2009).

Onsite visits on two separate occasions were completed during the study in an attempt to understand the operations at this facility and the effects of hydropower operations on downstream flows. The minimum flow release from the dam mandated by the State of North Carolina is 70 ft³/s, and it occurs through a gate opening that is locked in a fixed position in the dam. A stage sensor installed on the dam monitors the water level behind the dam and controls the ability to operate the turbine. When the lake level is 3 in. above the crest at the sensor, operation of the turbine begins to transition so that by the time the lake level is within 1 in. of a threshold at the top of the dam, the turbine has stopped and no further power generation occurs.

Power generation is not efficient when the discharge is less than 250 ft³/s as measured at the upstream streamgage on the Deep River near Moncure (site 3; William D. Brooks, owner, Lockville hydropower station, oral commun., November 9, 2009). On the basis of a recent streamflow analysis completed for this streamgage, this discharge is between the 70th and 75th exceedance percentiles of daily mean discharges, which implies that power generation is limited at or below 25 to 30 percent of the lowest daily discharges recorded at the Moncure streamgage during the 1930–2010 water years. No formal discharge-turbine rating is available; however, some general ranges were obtained during the second onsite visit. Approximately 140 kW of power is generated at about 250 ft³/s of discharge through the turbine. The owner of the power plant states that the most power that can be efficiently generated is about 600 kW. This level of power generation requires about a 550 ft³/s discharge through the turbine. Thus, under the current operation of one turbine, the discharge diverted through the canal between the dam and powerhouse during maximum power generation is between 500 and 600 ft³/s.

The discharge release from the turbine into the Deep River is subject to backwater effects caused by storage upstream from Buckhorn Dam and (or) large flow releases from Jordan Lake Dam, which also create backwater conditions on the Deep River from its mouth to the hydropower operation. To meet FERC regulations, hydropower operations must be completed in a run-of-river manner with no peaking cycles permitted.

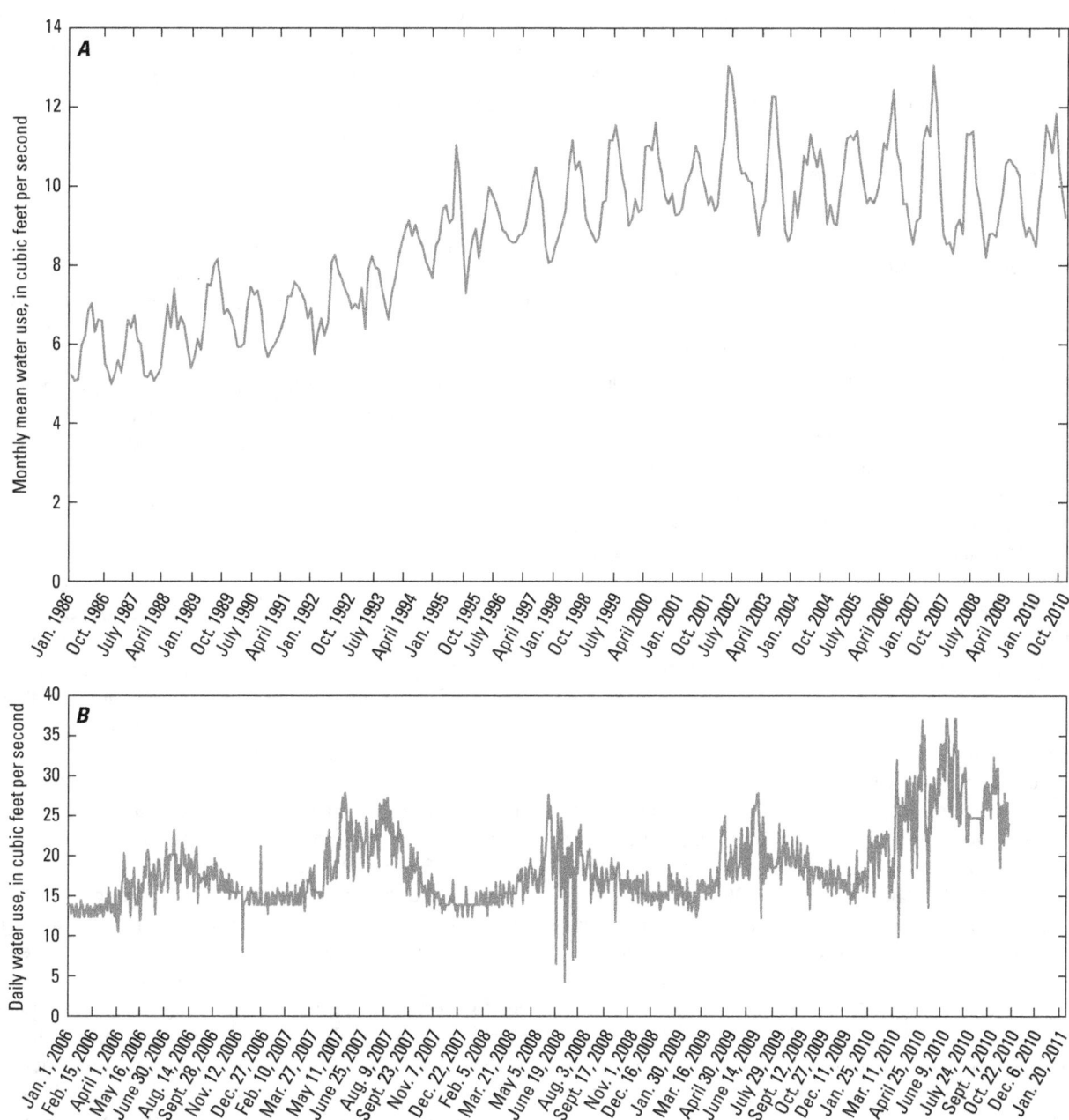

Figure 26. Water use reported for (*A*) municipal facility F9 for 1986–2010, and (*B*) municipal facility F10 for 2006–2010 in the Cape Fear River study area, North Carolina.

Records maintained to document the water level at the dam (as detected by a continuous sensor) and amount of power generated are not required to be submitted to FERC or any other regulatory agency. Records were not made available for the full study period but were made available for selected dates (William Brooks, owner, Lockville hydropower station, oral commun., January 27, 2011).

During the series of six synoptic measurements completed in 2009, hydropower generation only occurred on two dates (April 27 and July 22). No power generation occurred during the remaining four synoptic dates (June 25, August 18, September 9, and October 1). Records indicate power generation

occurred all day on April 27 and during limited hours on July 22. Instantaneous discharges on April 27 at the upstream streamgage on the Deep River (site 3) were between 400 and 500 ft³/s. On July 22, the turbine was operated between the hours of 1300 and 1600 until the water level at the sensor on the dam declined to less than 1 in. above the top of the dam, resulting in the shutdown of the turbine. The instantaneous discharge at the streamgage was steady at 162 ft³/s for the 3 hours of operation on this date. This supports the assumption that power generation is inefficient below a discharge of 250 ft³/s, but further data are needed to confirm this conclusion. While no numerical flow-routing models were completed as part of this study to confirm the effects of hydropower generation, the release of discharge into a backwater-affected reach under presumed run-of-river operations does not appear to be a major factor behind flow losses in the study reach.

Regional Power-Plant Diversion and Flow through Diversion Canal

The largest water user in the study area is a regional power producer that generates electricity at two separate facilities. The first is a coal-fired power-generation plant immediately adjacent to the Cape Fear River just downstream from the confluence of the Haw and Deep Rivers (fig. 1, site F7). As previously discussed, the second facility is the Shearon Harris Plant adjacent to Shearon Harris Lake, an impoundment of Buckhorn Creek, the second largest tributary in the study area. Because of available continuous discharge records at the streamgage on Buckhorn Creek near Corinth (site 20) downstream from Shearon Harris Lake Dam, a detailed assessment of water-use patterns at this facility was deemed unnecessary for the objectives of this study.

At the first facility operated by the regional power producer, water is withdrawn from the Cape Fear River and passed through the plant for cooling purposes before being discharged into the diversion canal adjacent to the main stem. After discharge from the plant, the water is passed through cooling towers as needed during June through September to reduce the temperature prior to final release to the canal. From the plant to the canal mouth immediately upstream from Buckhorn Dam, the diversion canal is 6.3 mi long and drains about 10.3 mi^2 of the study area (site 16). The dam, which was constructed in 1908 (Ragland and others, 2003), has provided a storage of water for the cooling operations and was actively regulated until December 1962 when it changed to a run-of-river mode of operation. Buckhorn Dam is approximately 5.75 mi downstream from the intake for the power plant.

The diversion canal was extended in 1954, and the added distance to its current mouth above Buckhorn Dam allows additional cooling time of the heated water released from the plant. The original canal mouth is approximately 1.75 mi downstream from the intake on the main stem, making the original canal distance about 2.25 mi.

The regional power producer compiles and submits discharge monitoring report (DMR) data to State regulatory agencies to track daily water withdrawals. The DMR data for the period April 2008 to December 2010 were compiled and coupled with concurrent daily discharge records at the continuous-record streamgage on the canal (site 8) in an effort to determine estimates of water consumption between the intake and streamgage. Estimated differences between the intake and canal discharge data typically were in the range of 30 to 50 ft^3/s (20 to about 30 Mgal/d) with values as high as 75 to 100 ft^3/s (about 50 to 65 Mgal/d). These ranges are much higher than the approximately 13.4 ft^3/s (8.65 Mgal/d) estimate in water loss from the cooling towers (Robin Bryson, Progress Energy Carolinas, written commun., November 30, 2009).

The regional power producer expressed concern that these comparisons are limited because of the methods used to calculate the daily withdrawals. Because the pump intake logs are used to estimate the flow discharged into the diversion canal, these estimates can only provide a daily "snapshot" of the flow volume through the plant and should not be used for making daily comparisons with the continuous discharge records observed at the canal streamgage (site 8; Robin Bryson, Progress Energy Carolinas, written commun., August 26, 2011).

Uncertainty in the amount of water consumption during the study prompted the regional power producer to undertake measures to adjust and validate pump curves used for estimating withdrawal amounts. These measures were completed in September 2010, and adjusted pump curves were applied to DMR data beginning in December 2010 near the end of the investigation. The regional power producer reported that percentage change between the old and newly adjusted pump flows among the varying combinations of pumps ranged from -7 percent to almost 43 percent (average 17.6 percent). Using the adjusted pump curves, the regional power producer also provided new estimates of the withdrawal volumes (at 15-minute intervals) for April 2008 through December 2010 (Robin Bryson, Progress Energy Carolinas, written commun., August 30, 2011). These estimates were used to calculate daily average withdrawal amounts for a second comparison with daily mean discharges at the canal streamgage (site 8).

The second comparisons between the withdrawal data for the period of interest (based on the adjusted pump curves) and the discharge records at the canal streamgage (site 8) indicated a smaller range, typically from 10 to 40 ft^3/s (6.5 to 25.8 Mgal/d) with some amounts exceeding 50 ft^3/s (32.3 Mgal/d) on some days (fig. 25B). A large part of this range still exceeds the approximately 13.4 ft^3/s (8.65 Mgal/d) estimate in water loss from the cooling towers provided by the regional power producer. The regional power producer continued to express concern about the utility of these comparisons to estimate water loss from the facility even after the pump curve adjustments were made (Robin Bryson, Progress Energy Carolinas, written commun., August 30, 2011). Thus, uncertainty surrounding reasonable estimates of consumption remained in effect at the end of the study. Further resolution of this issue requires the collection of additional data after December 2010. Following the investigation, it was learned the power company is tentatively planning to close the regional power plant by mid-2013 because of changing environmental regulations and economic conditions (Mick Greeson, Progress Energy Carolinas, written commun., September 7, 2011).

Evaporative Losses

Evaporation is part of the hydrologic cycle (U.S. Geological Survey, 2011b), and it is understood that part of the flow loss from rivers is attributed to the removal of water from the earth's surface. Evaporation from the main stem was estimated during this study to understand the portion of flow losses attributed to this process. Evaporation analysis indicates

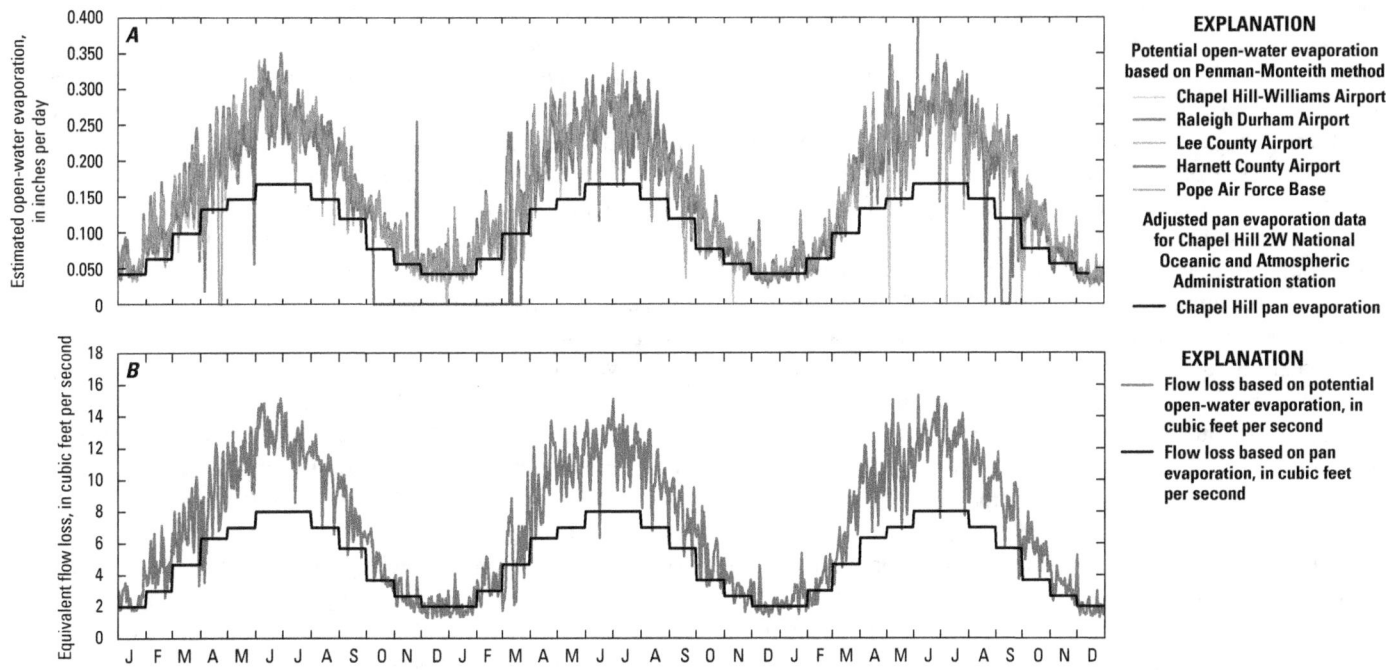

Figure 27. (*A*) Estimated open-water evaporation and (*B*) equivalent flow loss in the study reach from potential open-water evaporation based on the Penman-Monteith model at five nearby National Oceanic and Atmospheric Administration (NOAA) stations and pan-evaporation data for the Chapel Hill 2W NOAA station (State Climate Office of North Carolina, 2010).

that the estimated range in flow loss due to open-water evaporation from the main stem and the Deep River is between 10 and 38 percent of the 37 ft³/s median flow loss (table 3; flow comparison 3, daily discharge) during the months from May through October.

Data on evaporative losses were compiled through two approaches: (1) an analysis of available pan-evaporation data for the Chapel Hill 2W NOAA weather station in Orange County and (2) a compilation of estimated open-water evaporation computed and compiled by the State Climate Office of North Carolina (2010) using the Penman-Monteith method used for computing reference crop evapotranspiration. Because of the intermittent nature of the data at the Chapel Hill 2W NOAA weather station, available records were sorted, and statistics were determined by month. The monthly median pan-evaporation amounts were adjusted by a pan coefficient of 0.7 (Farnsworth and Thompson, 1982). Adjusted monthly median evaporation on the basis of the Chapel Hill data ranged from 0.05 inch per day (in/d) during the winter to 0.15 in/d during the summer (fig. 27A). The monthly medians then were multiplied by the surface areas determined for the Haw, Deep, and Cape Fear Rivers in the study area to compute a volume of water loss (expressed as equivalent daily flow in cubic feet per second) due to estimated open-water evaporation (fig. 27B) during the 3 years (2008–2010) corresponding to the study period. Equivalent daily flow loss from the main stem and the Deep River on the basis of the pan-evaporation data ranged from about 2–3 ft³/s during December through February to about 7–8 ft³/s during June through August (fig. 27B).

During the investigation, records of estimated daily open-water evaporation computed using the Penman-Monteith method were obtained from databases maintained by the State Climate Office of North Carolina (2010). This method estimates reference crop evapotranspiration rates for a well-watered reference surface on the basis of physical atmospheric observations of solar radiation, temperature, wind speed, and relative humidity. Daily records were accessed for five nearby NOAA stations in the vicinity of the study area: Chapel Hill-Williams Airport (NOAA station KIGX), Harnett County Airport (KHRJ), Lee County Airport (KTTA), Pope Air Force Base (KPOB), and Raleigh-Durham International Airport (KRDU, fig. 8). The daily open-water evaporation estimates among the five sites using the Penman-Monteith method were averaged to determine a regional evaporation amount. Daily evaporation based on the Penman-Monteith data ranged from about 0.06 in/d during the winter to 0.25 and 0.3 in/d during the summer (fig. 27A). As with the pan-evaporation method, the averaged values were then multiplied by the surface area determined for the main stem and the Deep River to compute a volume of water loss. The equivalent daily flow-loss estimates on the basis of the Penman-Monteith method ranged from about 2 ft³/s during December through February to about 11–14 ft³/s during June through August (fig. 27B).

Comparison of the daily flow losses indicates substantial differences between the two methods for computing evaporation, particularly during the summer months. Discussion of the differences between the methods is beyond the scope of this report; however, the differences during the summer months highlight the uncertainty and complexities associated with estimating the portion of flow loss due to evaporation from the study reach. Part of the uncertainty is due to the recognition that no pan-evaporation data have been collected within the study area. Pan coefficients used to adjust pan-evaporation data vary by season during the course of a year (Yonts and others, 1973) and vary on the basis of the type of water body (river as opposed to lake with large fetch) for which evaporation estimates are needed (David I. Stannard, U.S. Geological Survey, oral commun., November 9, 2011). Evaporation estimates based on the Penman-Monteith method for locations far from water bodies tend to be higher than actual evaporation, because of the different meteorological measures used in the method, than would be measured closer to the water surfaces (David I. Stannard, U.S. Geological Survey, oral commun., November 9, 2011). The NOAA stations for which Penman-Monteith estimates were compiled were all located at airport facilities as opposed to nearby water bodies. For the purposes of this study, both methods were considered, by estimating the potential flow loss from the study reach due to evaporation from the main stem and the Deep River, to be in the range of 4 to 14 ft^3/s during May through October (fig. 27B). The flow-loss analyses completed during this study indicate that 82 percent of flow-loss days occurred during these months. The estimated range in flow loss due to open-water evaporation is equivalent to 10 to 38 percent of the 37 ft^3/s median flow loss (table 3; flow comparison 3, daily discharge).

Comparisons of Water Use, Evaporation, and Flow-Loss Occurrences during 2008–2010

Comparisons were made of daily water-use diversions and evaporation losses with flow-loss occurrences during the period April 2008 through September 2010. This period was chosen because of its concurrence with the period of record at the streamgage on the adjacent diversion canal (table 1, site 8). The comparison was completed to assess how the magnitude of total losses due to water consumption and evaporation compared to calculated flow losses.

During the period of interest, flow analyses indicated 289 days of flow-loss occurrences at the streamgage on Cape Fear River at Lillington (site 28). Filtered daily flow losses occurred on 34 days during the same period (filtered for days of steady flow conditions at the upstream inputs relative to the previous day). Flow losses ranged from 0.5 ft^3/s (September 22, 2010) to about 140 ft^3/s (February 21, 2010) with a median flow loss of 20.7 ft^3/s. Although the maximum flow loss met the criteria applied to filtered flow analyses, the flow loss actually is not a true flow loss because it occurred 2 days after an adjusted flow release of 4,000 ft^3/s at Jordan Lake Dam on February 19, 2010.

Among the 34 days of filtered flow losses, the total water-use diversions for the five industries and two municipalities exceeded the computed flow losses on 26 days. With evaporative losses added, the total losses exceeded the computed flow

losses at the streamgage on 1 and 2 additional days, depending on whether evaporative losses are based on the pan-evaporation data or the Penman-Monteith computations, respectively. For the remaining 6 days when total losses from water use and evaporation did not exceed the computed flow losses at the streamgage, flow losses could be attributed to the addition of water consumption from the regional power plant (facility F7) on 2 other days during the period of interest. Among these 6 days, the balance of unaccounted flow loss after adjustments for water-use diversions and evaporative losses ranged from about 7 to 104 ft^3/s, less than 16 ft^3/s on 4 of the 6 remaining days.

Water consumption from the power plant could not be determined reliably (based on the data provided from the facility) during the study as previously discussed. However, water consumption from the facility could account for part of the flow losses on at least 3 of the 6 remaining days, which occurred during August 2009 at a time when the water consumption rates from the facility can be among the highest during the year. The remaining 3 days were during the winter and early spring following the passage of higher flows along the main stem and likely a peak flow passing through the system.

In comparing the surface-water, water-use, and evaporation data compiled for 2008–2010, it is evident that documented water diversions combined with losses from open-water evaporation can exceed the net flow gain in the study area. The median flow loss was determined to be 37 ft^3/s (table 3; flow comparison 3, daily discharge), which is slightly greater than the combined municipal water withdrawals of 23 to 36 ft^3/s. The effects of open-water evaporation from the main stem and the Deep River during July through August equals between 4 and 14 ft^3/s of equivalent flow loss removed from the hydrologic system. Summing the two ranges of water diversions and evaporation provides a total range of 27 to 50 ft^3/s of equivalent flow loss. The small magnitude of measured discharges from the four smaller tributaries during the series of synoptic measurements also confirm that net gain in flow along the study reach is limited, particularly under base-flow conditions during warmer months.

Effects of Diversion Canal Flows on Cape Fear River at Lillington

The operation of the streamgage on the diversion canal (fig. 6, site 8) downstream from the regional power plant (facility F7) provided additional insight into the occurrence of an apparent flow loss at the streamgage at Lillington (fig. 7, site 28). As previously discussed, the diversion canal travels adjacent to the Cape Fear River allowing heated waters from the power plant sufficient time to cool before merging into the main stem.

In the years leading up to the investigation, sudden decreases in the instantaneous streamflow would occasionally be observed in the Cape Fear River at Lillington (site 28), often prompting an inquiry by the USACE about the rapid changes. During the latter stages of the investigation, a connection was established between the occasional sudden

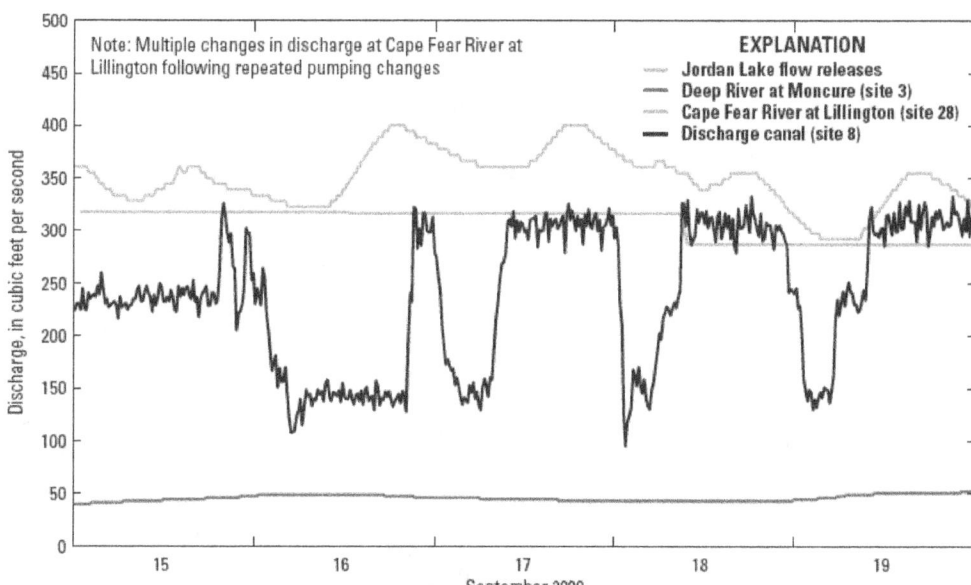

Note: Multiple changes in discharge at Cape Fear River at Lillington following repeated pumping changes

EXPLANATION
— Jordan Lake flow releases
— Deep River at Moncure (site 3)
— Cape Fear River at Lillington (site 28)
— Discharge canal (site 8)

Figure 28. Effect of changes in instantaneous discharges at diversion canal streamgage (site 8) and Cape Fear River at Lillington (site 28), North Carolina, on September 15–19, 2009.

changes in canal flow and downstream flows at the Lillington streamgage. From time to time, some or all of the pumps used to withdraw water into the regional power plant (facility F7) are stopped for durations ranging from one-half day to about 2 days, mainly based on the run times of the generating units in response to power demands. Changes due to pumping adjustments can be observed in the discharge records at the streamgage on the diversion canal (site 8) downstream from the power plant.

An assessment of the daily discharges and subsequent inspection of the instantaneous-value hydrographs for the canal streamgage (site 8) indicate at least 24 instances from the start of data collection in April 2008 through December 2010 during which the flows suddenly decreased, and then increased shortly thereafter, by magnitudes of 100 to more than 200 ft^3/s. Flows at the Lillington streamgage (site 28) during 14 of these instances were in the high-flow range above 1,000 ft^3/s. Thus, changes in the canal flow were "masked" by the larger flows in the main stem. Flows at the Lillington streamgage during the remaining 10 instances, however, were at or below 600 ft^3/s. At these discharges, the above-referenced changes in the canal discharges were determined to affect the pattern of discharges at the Lillington streamgage.

The scope of this investigation did not include flow routing or model development that would permit in-depth assessment of the effects of starting or stopping main-stem withdrawals and subsequent canal discharges on main-stem flows downstream. For the 10 instances during the period of interest when effects were noted, however, visual inspection of the instantaneous hydrographs permitted a qualitative assessment of this cause-and-effect relation. The changes in flows are illustrated in 4 of the 10 instances (figs. 28–31).

Multiple oscillations in the discharges at the Cape Fear River at Lillington (site 28) during September 15–19, 2009, echo the pattern of adjusted discharges in the diversion canal for the same period (fig. 28). Little to no changes in discharge

occurred in the flow releases from Jordan Lake Dam (site D1) or at the streamgage at Deep River at Moncure (site 3) during the period. Variations in canal discharge ranged from 150 to 200 ft^3/s with decreases typically occurring over a period of a few hours and increases occurring over a slightly longer period. The decreases and increases in discharge at the Lillington streamgage ranged from 40 to 60 ft^3/s. The hydrograph for this period indicates that flows at the Lillington streamgage increased on a time-lagged basis when the discharge through the diversion canal substantially decreased. In a similar manner, discharges at the Lillington streamgage appear to decrease in response to a corresponding increase in discharge through the diversion canal.

The magnitude and duration of discharge changes at the Lillington streamgage (site 28) are governed by the magnitude and timing of changes in canal discharge (fig. 28). Hydrographs for three other selected dates (figs. 29–31) confirm this pattern. On November 1, 2009, the canal discharge decreased during the morning by approximately 150–200 ft^3/s, which resulted in an increase in discharge of about 60 ft^3/s at the Lillington streamgage beginning about 8 hours later (fig. 29). When the canal discharge increased on the following midnight, however, the effects on discharge at the Lillington streamgage are not as evident (or are indeterminate) given that discharges were already in a declining pattern beginning at hour 2200 on November 1. The hydrograph for this date also indicates the discharges at Deep River at Moncure (site 3) were in recession from October 31 into the early morning hours of November 1 (fig. 29).

On October 28, 2010, the canal discharge decreased during the morning by approximately 250 ft^3/s, which resulted in an increase in discharge of about 90 ft^3/s at the Lillington streamgage about 8 hours later (fig. 30). Between noon and 1900, the canal discharge increased about 350 ft^3/s. At about 2030 (about 8 hours after the initial increase), the discharge at the Lillington streamgage declined about 150 ft^3/s over a

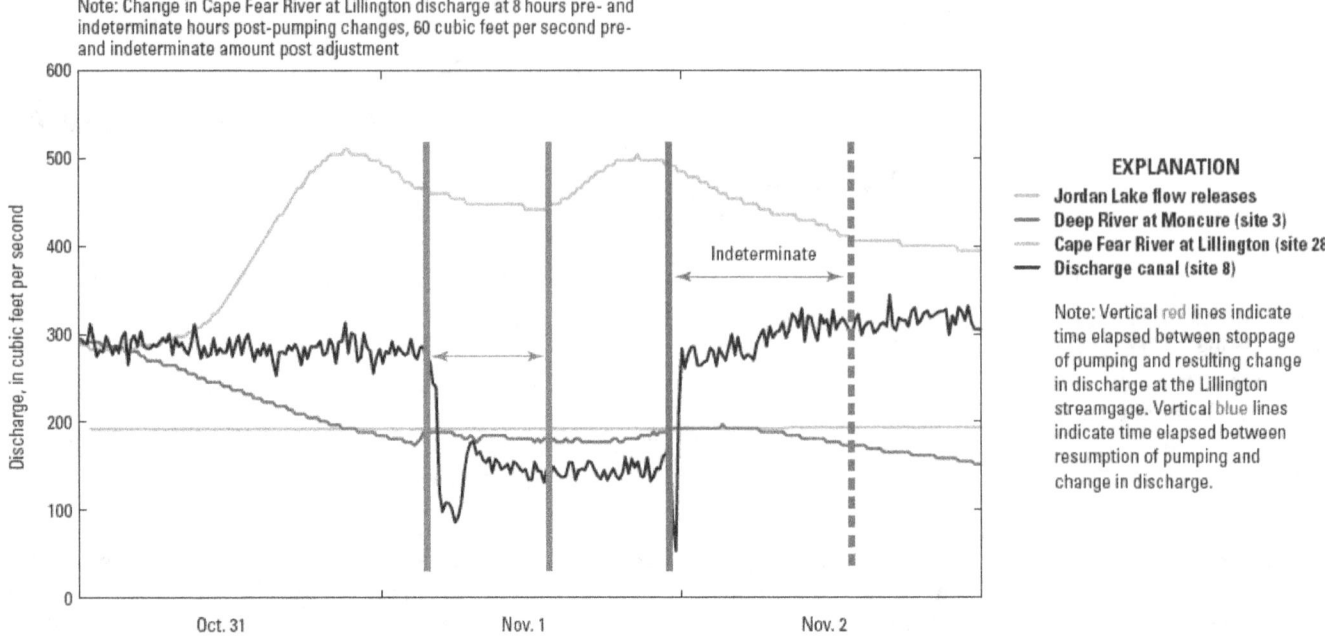

Figure 29.   Effect of changes in instantaneous discharges at diversion canal streamgage (site 8) and Cape Fear River at Lillington (site 28), North Carolina, on November 1, 2009.

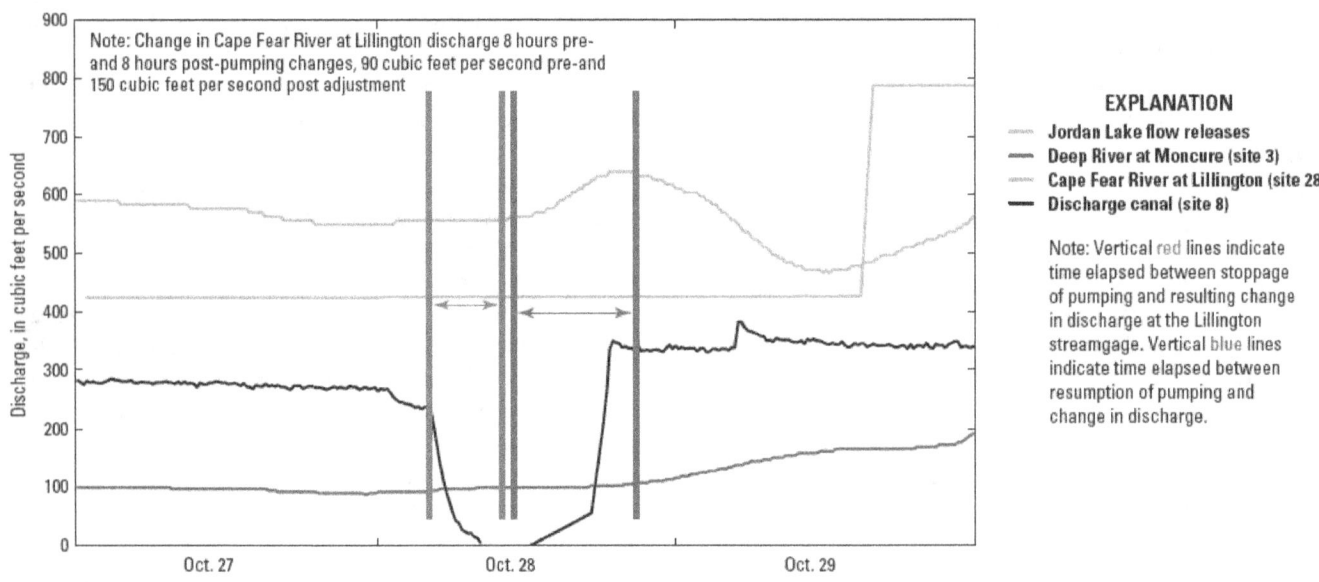

Figure 30.   Effect of changes in instantaneous discharges at diversion canal streamgage (site 8) and Cape Fear River at Lillington (site 28), North Carolina, on October 28, 2010.

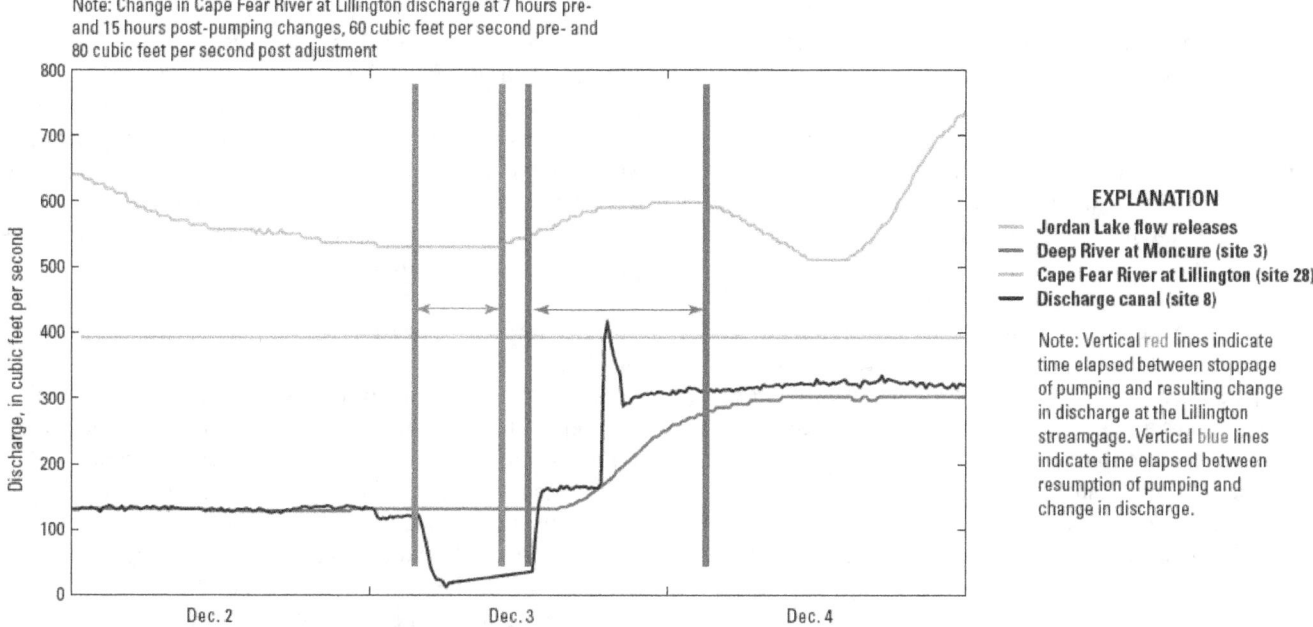

Figure 31. Effect of changes in instantaneous discharges at diversion canal streamgage (site 8) and Cape Fear River at Lillington (site 28), North Carolina, on December 3, 2010.

period of 16 hours. The hydrograph for this date also indicates little to no changes in the discharge at Deep River at Moncure (site 3) or in flow releases from Jordan Lake Dam (fig. 30).

The hydrograph for December 3, 2010, indicates the canal discharge decreased by approximately 100 ft³/s during the morning, which resulted in an increase in discharge of about 60 ft³/s at the Lillington streamgage about 7 hours later (fig. 31). Beginning at 1230, the canal discharge increased about 375 ft³/s. At about 0230 (approximately 15 hours after the initial increase), the discharge at the Lillington streamgage declined about 80 ft³/s over a period of 15 hours. The hydrograph for this date also indicates little to no changes in the discharge at Deep River at Moncure (site 3) or in flow releases from Jordan Lake Dam (fig. 31) through the period during which the canal discharge decreased then increased again. However, beginning at about 1500 on December 3, the discharge at the Deep River streamgage began to increase, and the discharge at the Lillington streamgage correspondingly began to rise almost 24 hours later on December 4 (fig. 31).

These hydrographs indicate that when discharge through the canal is decreased or stopped for a short period, that amount of discharge is immediately available for travel in the Cape Fear River, which results in an increase in discharge at the downstream Lillington streamgage (site 28) beginning about 8 hours later after the main-stem channel storage has been satisfied. When discharge from the Cape Fear River is subsequently diverted back into the canal, there usually is (but not always) a time-lagged decrease on the downstream

discharges at the Lillington streamgage beginning about 8 to 16 hours later. Part of the longer time lag may be due in part to the time to refill (satisfy storage in) the diversion canal. As observed by USGS field personnel, the channel was nearly empty when the discharge substantially decreased in response to a stoppage in pumping operations. Another factor affecting the lag time may be the interaction of flows and, consequently, travel time between the diversion and the main stem affected by the storage immediately upstream from Buckhorn Dam.

Groundwater and Surface-Water Interaction

Rivers gain water from or lose water to the underlying aquifer depending on climate, geologic setting, changes in streambed slope, meanders in the stream, and hydrologic gradients between the river and the aquifer. As part of this study, several complementary methods were applied to assess conditions of gain or loss through groundwater and surface-water interaction within regional (large scale) and local (small scale) confines of the Cape Fear River. Regional groundwater characteristics over the entire 24.5-mi river study reach were assessed using an EM-induction geophysical survey of the channel bottom to identify possible locations of diabase dikes. The EM-induction survey enabled selection of a suitable location for deployment of a FO-DTS survey that measured local changes in the temperature of riverbed sediments near exposed diabase dikes. A regional aerial flyover of the Cape Fear and Haw Rivers was conducted using infrared techniques to detect regional ground-water discharge by measuring variations in water temperature near the flood-plain sediments. Additionally, three reaches of the Cape Fear River were selected for local groundwater-level monitoring using a transect of piezometers installed in the flood plain to determine the validity of the underlying assumption that the groundwater-flow system near the Cape Fear River is functioning in "gaining-stream" conditions. These groundwater-measurement methods differ in resolution and time scale from the methods applied to the surface-water data collection. The groundwater data-collection methods applied to the subsurface yield point estimates at a small scale. The groundwater data collected during this investigation indicate the possibility of localized flow loss during the summer, particularly in the reach above Buckhorn Dam. However, there was no indication of unusual patterns that would cause substantial flow loss from the study reach as a result of groundwater and surface-water interaction at the river bottom.

In humid regions, such as those present in North Carolina, a typical river receives groundwater discharge by the seepage of groundwater into the riverbanks or riverbed (base flow; Winter and others, 1998; Weaver and Pope, 2001; LeGrand, 2004). Under these gaining-stream conditions, the groundwater level near the river is higher than the water level in the stream channel, which causes stream discharge to increase in a downstream direction even if no tributaries are present. If the river is a "losing stream," the groundwater level near the river is below the river level. Streams can be gaining streams along one river reach and losing streams along another reach, depending on the relation to the water table.

Under conditions of drought, storm events, abrupt changes in channel gradient (both natural and those caused by manmade diversions), nearby groundwater pumping, evaporation, and transpiration of groundwater by vegetation, a river reach can alternate between losing and gaining on a daily or seasonal basis. During storm events, water can overflow the river onto the streambank as the river stage rises causing surface water to infiltrate into the groundwater beneath the flood plain and resulting in temporary localized changes to the water-table altitude. These differences in flow dynamics and variations in underlying geologic units can influence regional and local groundwater interaction.

Electromagnetic Geophysical Survey

A geophysical survey of the 24.5-mi river bottom in the study reach was conducted to identify possible locations of diabase dikes. A GEM–2 EM-induction geophysical instrument was used at base-flow conditions to map the presence of diabase dikes that may act as impermeable boundaries to groundwater discharge. Because of the higher iron and magnetite content in diabase dikes, compared to the surrounding geologic material, these geologic features beneath riverbed sediments can be mapped as higher variations in conductivity and magnetic susceptibility in the lower frequency EM response.

The majority of the detected anomalies in the geophysical survey results were located downstream from Buckhorn Dam (figs. 32–33) in the Carolina Terrane, Cary sequence rocks, while relatively few occurrences were detected upstream in the Mesozoic (Triassic) basin rocks. In the Haw and Cape Fear Rivers of the upper reach, the numerous conductive anomalies detected along the left side (when viewed in the downstream direction) likely resulted from variations in water-column thickness or instrument drift. Many anomalies were detected on both sides of the Haw River in the upper reach 1.75 mi downstream from Jordan Lake Dam and in areas where electric power-transmission lines spanning the river affected the instrument.

On the Cape Fear River, the depth penetration of the GEM–2 instrument may have been limited by deeper water-column depths in the main stem, particularly on the reach between State Highway 42 and Buckhorn Dam. In addition to more occurrences of increased magnetic anomalies in the lower reach, there also were more occurrences of increased signal detections along both sides of the Cape Fear River channel between Buckhorn Dam and the streamgage at Lillington (fig. 33, site 28). Examination of the mapped survey in the lower reach indicates increased magnetic signals on both sides of the channel.

Previous geologic mapping by Burt and others (1978) in central North Carolina covers the northwest part of the study area and indicates that diabase dikes typically strike northwest-southeast (fig. 32). Given that the orientation of the main stem is similar to these previously mapped dikes, it is inconclusive whether the conductive anomalies in the lower reach correspond to one or more diabase dikes in this reach. Although detailed geologic maps of diabase dike outcrops were not available for the lower reaches, multiple locations of mapped conductive anomalies from the EM survey indicate the potential presence of these relatively impermeable hydrologic boundaries, most notably between Buckhorn Dam and the confluence with Buckhorn Creek in Raven Rock State Park (including the group camp area near the downstream end of the park boundary) and in the reach abetween PZ2 and Lillington (fig. 33). Because of the clustering of conductive anomalies, presence of riffles, and relatively shallow river depths identified using the GEM2 EM-induction tool, the group camp area in Raven Rock State Park was selected as the location for the FO-DTS survey.

Groundwater Temperature

Two heat-tracer methods were employed to delineate groundwater discharge zones in the study area—a DTS survey of a 1-mi long reach of the Cape Fear River and a thermal infrared survey of the Haw and Cape Fear Rivers. Temperature traditionally has been shown to be an excellent tracer of groundwater movement (Stonestrom and Constantz, 2003). Natural variations in streamwater temperature patterns are used to assess the interaction of river water with shallow groundwater. The temperatures of surface-water bodies are

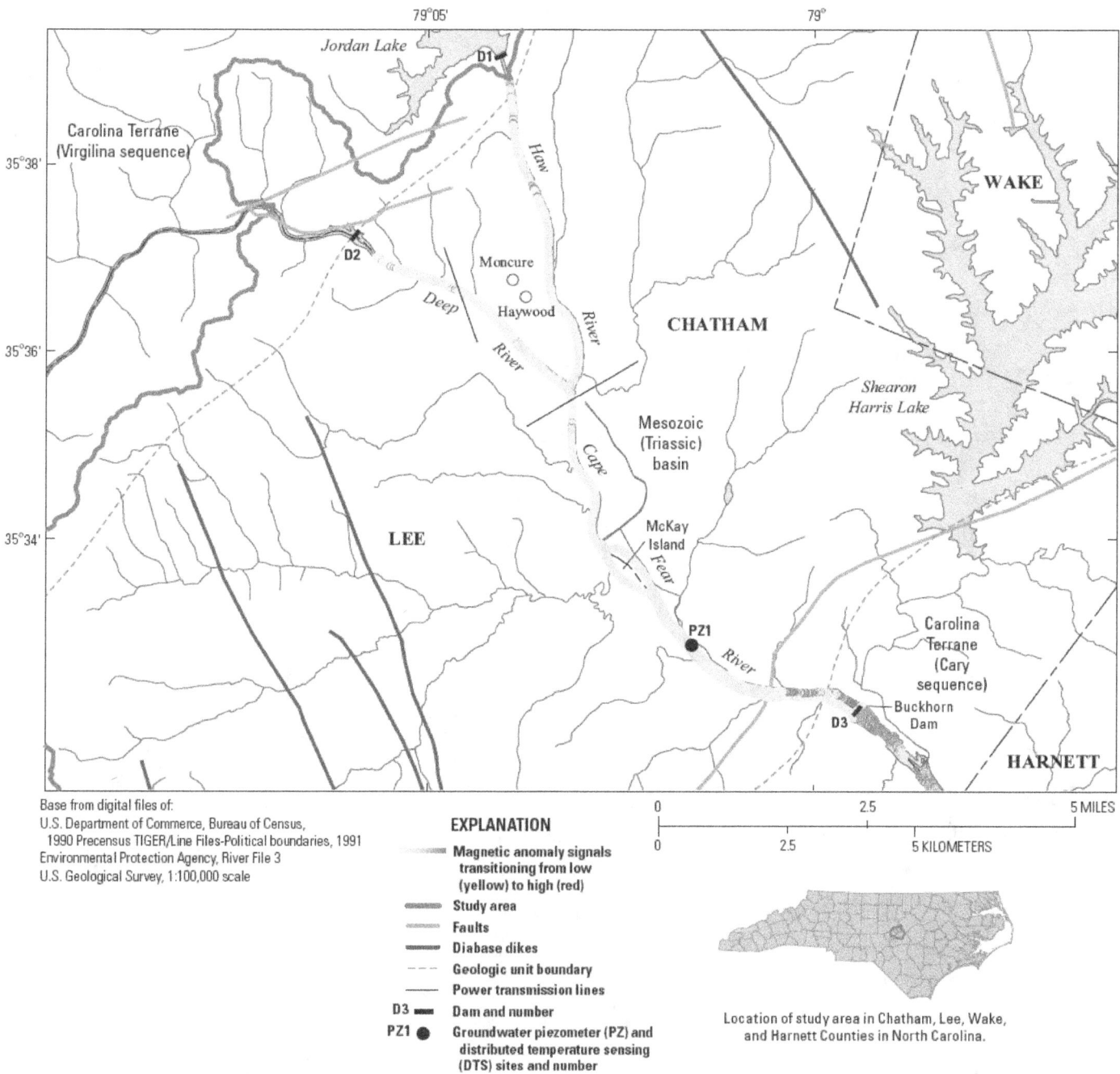

Figure 32.   Locations of increased magnetic anomalies in the upstream reach from Buckhorn Dam, North Carolina.

variable and change daily in response to seasonal and meteorological changes, whereas groundwater temperatures are relatively stable year round. Groundwater discharge into surface
water can be recognized as a cold signature in the summer and
a warm signature in the winter. These methods of measuring
differences in temperature can be used to identify areas of
groundwater discharge (gaining-stream conditions) into the
Cape Fear River but generally cannot distinguish between no-
flow and losing reaches.

Distributed Temperature Sensing (DTS) Survey

About 3,000 ft of fiber optic cable was deployed on the
bed of the Cape Fear River at the Raven Rock State Park reach
(fig. 33, site DTS) to evaluate if potential effects of diabase
dikes, delineated from the EM geophysical survey, could be
preferentially directing groundwater discharge. Continuous
temperature measurements were made by the FO-DTS from
1800 hours on August 11 through 1000 hours on August 13,

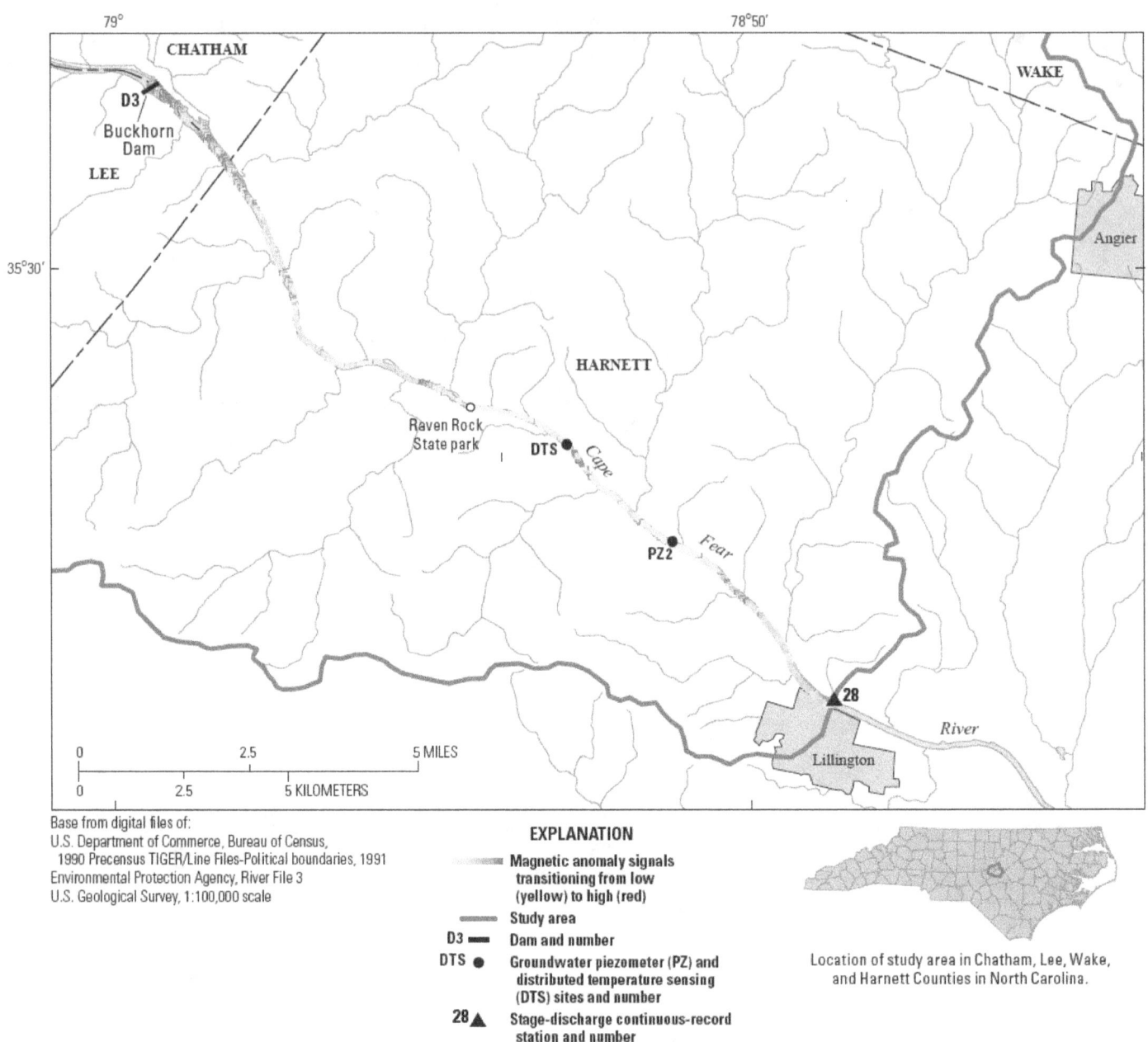

Figure 33.   Locations of increased magnetic anomalies in the downstream reach from Buckhorn Dam, North Carolina.

2009. The cable was laid in a U-shape pattern along the bottom of the riverbed, paralleling the right riverbank (fig. 34). At the Raven Rock State Park reach, the riverbed was composed of exposed bedrock and was devoid of sediment, with the exception of a limited quantity of sand-sized particles that accumulated around boulders. The cable crossed three large shallowly submerged quartz-boulder shoals and six conductive anomalies identified during the EM geophysical survey. The stage of the Cape Fear River at the Lillington streamgage (site 28) was low and stable (base-flow conditions) and fluctuated less than 0.03 ft over the 3-day survey with a daily mean discharge from

469 to 486 ft³/s. Water temperatures measured near the riverbed during the survey ranged from about 28 to 34 °C.

The FO-DTS data are displayed as colored thermal transects in figure 34 at four different time periods on August 12. No temperature anomalies of colder water were measured during the survey, which indicates that at the time of the survey, the Cape Fear River was a no-flow or losing stream in that particular river reach. If groundwater discharge were entering the Cape Fear River through riverbed seepage, a persistent cold temperature anomaly (about 15 to 20 °C) would have been identified.

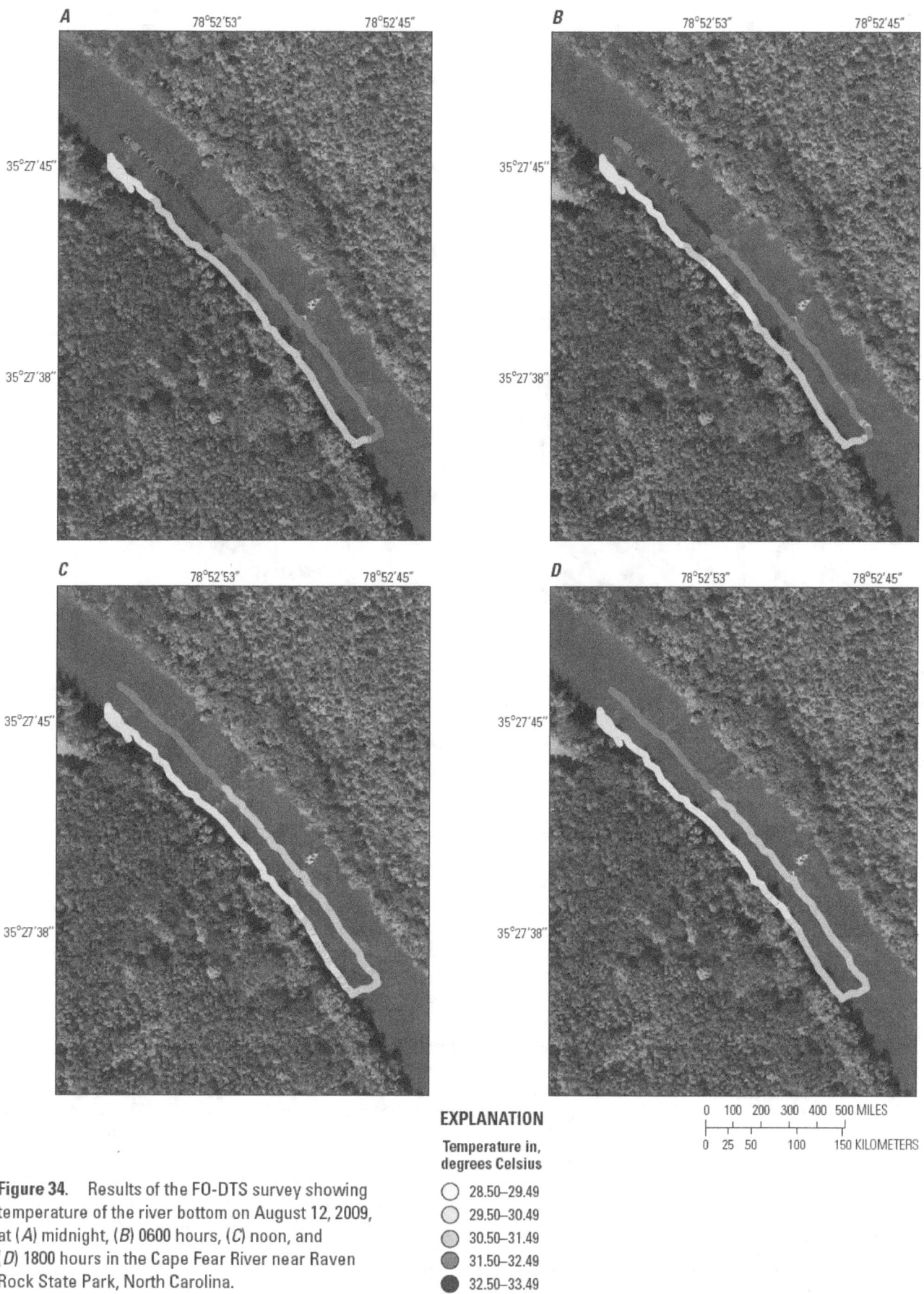

Figure 34. Results of the FO-DTS survey showing temperature of the river bottom on August 12, 2009, at (*A*) midnight, (*B*) 0600 hours, (*C*) noon, and (*D*) 1800 hours in the Cape Fear River near Raven Rock State Park, North Carolina.

EXPLANATION

Temperature in, degrees Celsius

○ 28.50–29.49
○ 29.50–30.49
◓ 30.50–31.49
◒ 31.50–32.49
● 32.50–33.49

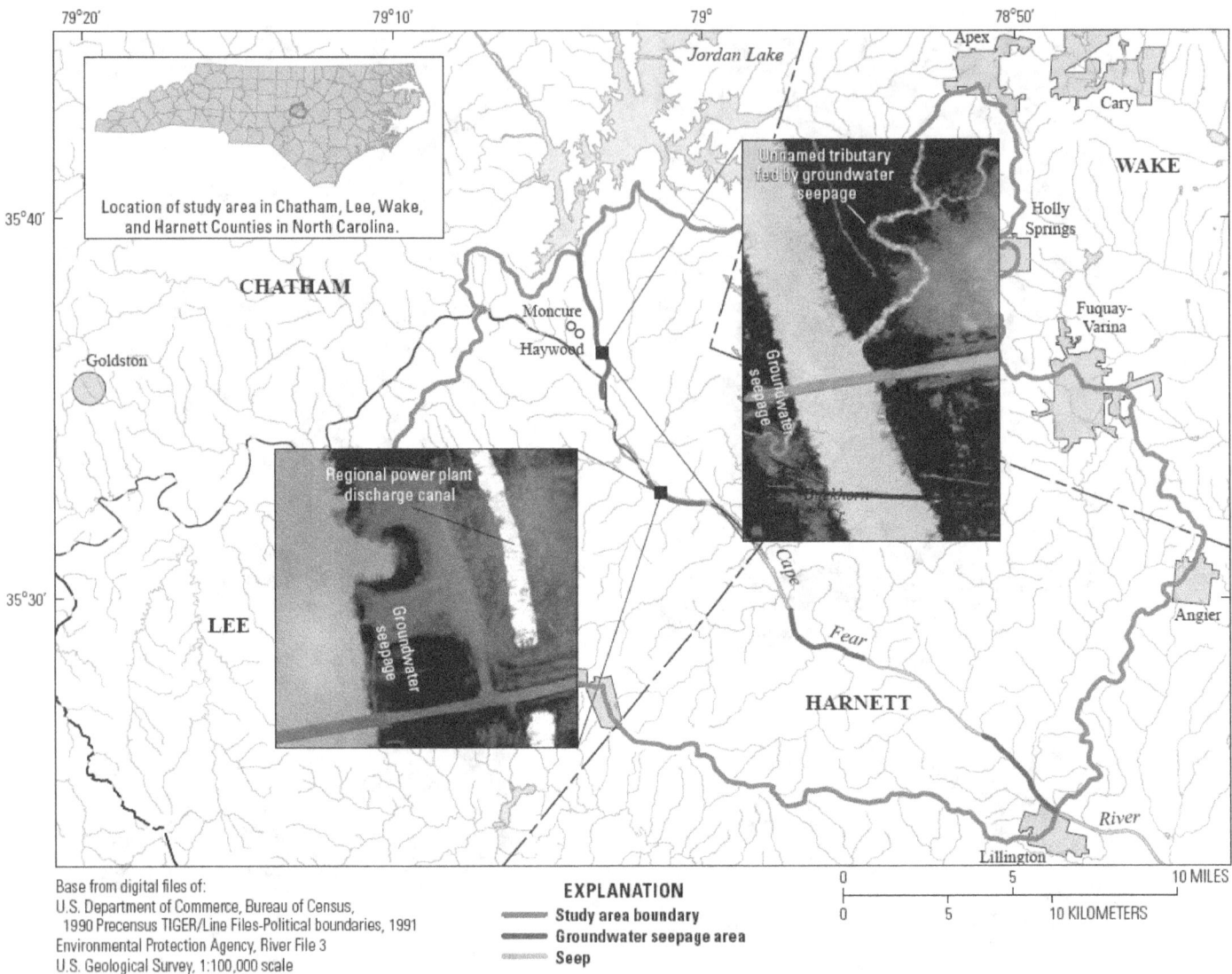

Figure 35.   Areas of groundwater discharge to the Cape Fear River between Jordan Lake Dam and Lillington, North Carolina, identified by thermal infrared imaging, February 27, 2010.

It should be noted that the lack of observed groundwater seepage during the survey does not imply that this reach is a no-flow or losing stream at all times during the year. It does suggest that, under conditions similar to those present during the survey (high ambient temperature, low river stage, high ET processes, and low flow), this reach is not a gaining stream. Additionally, the FO-DTS survey was unable to discern losing-stream conditions, because the fiber-optic cable could not be deployed vertically beneath the river to measure temperature gradient with depth. The presence of competent bedrock instead of riverbed sediments lining the channel made vertical temperature measurements impossible in this reach.

Thermal Infrared (TIR) Imaging

A regional TIR aerial flyover of the Cape Fear and Haw Rivers was conducted on February 27, 2010, to detect regional groundwater discharge by qualitatively measuring variations in water temperature near the flood-plain sediments. The application of remote temperature sensing in rivers is a relatively recent development that has proven effective for assessing stream temperature patterns (Banks and others, 1996; Belknap and Naiman, 1998; Torgersen and others, 2001). Conventional methods of in-stream temperature measurement by placing data recorders in the stream are useful in evaluating groundwater and surface-water interaction but generally are spatially limited (Stonestrom and Constantz, 2003). The spatial data needed to map sources of thermal heterogeneity at the large watershed scale are facilitated by airborne TIR imaging. With TIR, differences in temperature can be used to qualitatively identify areas of groundwater discharge (gaining-stream conditions) into the Cape Fear River but generally cannot distinguish between no-flow and losing reaches. Areas interpreted from the TIR images as groundwater seepage (gaining stream) are shown in figure 35.

The TIR survey conducted as part of this study proved highly effective for qualitatively examining the spatial distribution of stream temperature. Four areas of groundwater discharge were present in the study area—between Jordan Lake Dam and the confluence of the Haw and Deep Rivers, between McKay Island and Buckhorn Dam, above Raven Rock State Park, and above the city of Lillington. Groundwater discharge was noted generally as diffuse groundwater seepage, which displays in TIR imagery as warmer temperatures present on the surface water across an area close to the riverbank. In two locations just downstream from Raven Rock State Park, discrete points of groundwater discharge (springs) were detected. Two closeups of TIR images that display the warmer groundwater seepage are shown in figure 35. Of immediate notice in one of the TIR images is the temperature contrast between the flows in the main stem and the adjacent diversion canal, a reflection of the much warmer water temperatures in the canal as it flows downstream from the regional power plant (fig. 35).

The TIR imagery collected as part of this study represents hydrologic conditions present at one point in time and does not represent conditions throughout the entire year. The TIR imagery does indicate, however, that under some hydrologic conditions during the winter, much of the study area is gaining an undetermined quantity of water through groundwater seepage in selected reaches along the main stem.

Hydrogeology

Groundwater availability and monitoring were not considered for much of the regional study area because of access restrictions and limited availability of long-term monitoring wells. Instead, three reaches of the Cape Fear River were selected for local groundwater monitoring using a transect of piezometers installed in the flood plain (locations PZ1, DTS, and PZ2; figs. 6, 7). The groundwater investigation focused on the calculation of local hydraulic gradients between the flood plain and the Cape Fear River to determine the validity of the underlying assumption that the groundwater-flow system near the Cape Fear River is functioning under gaining-stream conditions, as suggested by the LeGrand (2004) conceptual model.

The regional power plant reach is located in Chatham County about 8 mi downstream from Jordan Lake Dam (site PZ1, fig. 6). The river stage in this reach is highly controlled by backwater created by Buckhorn Dam, which is located about 2 mi downstream. The right riverbank (looking downstream) is a high, steeply cut bank (over 7 ft of exposed face) at typical base-flow stage, whereas the left bank gently slopes away from the river to a broad, flat flood plain. In the area of the regional power plant reach, the Cape Fear River is about 350 ft wide and the water depth at base flow exceeds 10 ft. A transect of piezometers was installed on the left bank perpendicular to the edge of the river to monitor horizontal and vertical groundwater flow through the flood-plain sediments. The arrangement and depth of piezometers are shown in figure 9A and construction details are listed in table 2. Three

piezometers were installed within the bed of the Cape Fear River (CH–231, CH -232, and CH–233) to monitor vertical groundwater gradients. Two piezometers were installed on the flood plain (CH–235 and CH–236) to monitor horizontal groundwater flow. One piezometer was installed in the bottom of the water column of the Cape Fear River (CH–234) to act as a reference for river stage.

The soil in the area of the regional power plant reach has been mapped as Riverview silt loam (Hayes, 2006), very deep, well-drained soil formed in alluvium on flood plains. Soil borings collected near the piezometers generally agree with this classification but indicate that soils in this location are slightly darker in color and finer in texture than typical Riverview silt loam (fig. 9A). The regional power plant reach was selected for monitoring because of the potential for a "losing reach" in this area resulting from the artificially high river stage raised by Buckhorn Dam.

The Raven Rock State Park reach is in Harnett County, about 8 mi downstream from Buckhorn Dam (fig. 7; site DTS). In this area, the width of the Cape Fear River can exceed 450 ft and is generally shallow, with water column depths less than 3 ft during base-flow conditions. Similar to the regional power plant reach, the channel wall on the right bank is high and steep (over 10 ft high at base-flow conditions), leading to a broad, flat flood plain that merges into steep upland slopes, whereas the left bank gently slopes away from the river into a flat flood plain. The Raven Rock State Park reach was selected because many magnetic anomalies were identified by the EM geophysical survey of the river in this area, potentially associated with diabase dikes that may serve as impermeable boundaries to groundwater flow (fig. 33).

An attempt was made to install a transect of piezometers at the Raven Rock State Park reach; however, several logistical complications prevented the installation. The bottom of the river in this location is devoid of sediment, and the exposed bedrock prevented the installation of piezometers in the river. Additionally, the right riverbank was inaccessibly high and steep, making it unsuitable for piezometer installation. While the left riverbank was accessible, the depth to bedrock was too shallow to install piezometers near the river. Ultimately, the intention to install a piezometer transect was abandoned in favor of conducting a thermal survey in this location.

The Bradley Road reach is in Harnett County about 3 mi upstream from the Lillington streamgage (fig. 7, site PZ2). Similar to the Raven Rock State Park reach, the right riverbank is high and steep (over 10 ft high at base-flow conditions) with an overlying flat flood-plain terrace. The left riverbank is more accessible as it slopes away from the river over several natural terraces to a broad, flat flood plain. In the area of the Bradley Road reach, the Cape Fear River is about 300 ft wide, and the water-column thickness at base flow is less than 10 ft. A transect of piezometers was installed on the left riverbank perpendicular to the edge of the river to monitor horizontal and vertical groundwater flow through the flood-plain sediments (fig. 9B); construction details are listed in table 2. Two piezometers were installed within the bed of the Cape Fear River (HR–060 and HR–061) to monitor vertical groundwater gradients. Three piezometers were installed on the flood plain (HR–063, HR–064, and HR–065) to monitor horizontal groundwater flow. One piezometer was installed in the water column of the Cape Fear River (HR–062) to act as a reference for river stage. Unfortunately, piezometer HR–061 was destroyed soon after installation during a storm in November 2009.

The soil in the area of the Bradley Road reach has been mapped as Chewacla loam (Spangler, 1994), a very deep, somewhat poorly drained soil formed in alluvium of Piedmont river valleys (fig. 9B). Soil borings collected near the piezometers generally agree with this classification but indicate that soils in this location are slightly more yellow in color and coarser in texture than typical Chewacla loam. The Bradley Road reach was selected for comparison to the regional power plant transect because it was likely representative of a typical gaining reach where diabase dikes were not delineated by the EM survey.

Continuous Monitoring

Groundwater levels and river-stage altitudes were monitored in piezometers installed at the regional power plant reach from October 2009 to November 2010 and at the Bradley Road reach from August 2009 to November 2010 to facilitate the calculation of hydraulic gradients between the flood plain and the Cape Fear River. Hydrographs of continuous water-level altitudes and river-stage elevations collected at the regional power plant and Bradley Road reaches are shown in figures 36, 37. Water-level altitudes measured in all piezometers were affected by their proximity to the Cape Fear River and rainfall events. Seasonal climatic trends typical of groundwater-level fluctuations across the southeastern U.S. Piedmont are not readily discernible as a result of surface-water releases from Jordan Lake (U.S. Geological Survey, 2011b).

Throughout the study period, water-level altitudes were variable between piezometers as a result of fluctuations in river stage and groundwater discharge, but distinctions can be noted when transect locations are compared. The Cape Fear River stage in the regional power plant reach fluctuated about 6 ft (fig. 36B) during the monitoring period of October 2009 to November 2010, while river stage in the Bradley Road reach, the fluctuation was about 15 ft (fig. 37B) during the monitoring period of August 2009 to November 2010. The smaller fluctuation in river stage in the regional power plant reach is a direct result of being located upstream from Buckhorn Dam where the fluctuation in stage is limited. At the regional power plant reach, all of the piezometers were intermittently inundated by the Cape Fear River for a few days during the period from November 2009 to February 2010 when the river stage exceeded the land-surface elevation of all the piezometers (fig. 36B). In contrast, most of the piezometers in the Bradley Road reach located about 11 mi below Buckhorn Dam were inundated by surface water from November 2009 to March

2010, with only brief periods when land surface was exposed at the piezometers (fig. 37B).

Changes in river stage had a noticeable effect on groundwater levels in both reaches. In the regional power plant reach, a decline in river stage resulted in a release of groundwater from bank storage and gaining-stream conditions several times in January, February, March, and April 2010 (fig. 36C). During periods of base flow, however, when the stage of the Cape Fear River is relatively stable and the air temperature is warm (greater than 70 °F), the flora populating the flood plain take up shallow groundwater through evapotranspiration (ET). Under these conditions in the regional power plant reach, the groundwater level in the flood plain quickly declined below the elevation of the river (losing-stream conditions). This effect is most noticeable in October 2009 and from June to September 2010 (fig. 36C). During the summer months in the regional power plant reach, it is not uncommon for groundwater altitudes measured in the flood-plain piezometers (fig. 36B, CH–235, CH–236) to be more than 3 ft lower than the Cape Fear River. In contrast, groundwater altitudes measured in piezometers at different depths in the bed of the Cape Fear River (fig. 36B, CH–231, CH–232) are rarely more than 1 ft lower than the river level (but still losing-stream conditions).

Changes in river stage had a similar effect on groundwater levels at the Bradley Road reach. A sudden decline in river stage resulted in a release of groundwater from bank storage and gaining-stream conditions several times in January, March, and April 2010 (figs. 37B, C). Unlike in regional power plant reach, periodic groundwater discharge (gaining-stream conditions) continued into May and June 2010. Similar to the regional power plant reach, when the stage of the Cape Fear River in the Bradley Road reach was relatively stable at base-flow conditions, the air temperature was warmer, and the flora populating the flood plain removed groundwater through ET, which resulted in decreased groundwater levels in the flood plain below the elevation of the river (losing-stream conditions). This effect was noticeable in August and September 2010 when temperatures were warmest and ET processes were dominant (fig. 37C). In the Bradley Road reach, groundwater altitudes measured in the flood-plain piezometers (fig. 37B, HR–063, HR–064, HR–065) were generally less than 1 ft lower than the river during the summer months, and groundwater altitudes measured in the streambed of the Cape Fear River (fig. 37B, HR–060) were rarely more than 0.25 ft lower than the river during the summer.

Water-level altitudes in the regional power plant and Bradley Road reaches were analyzed over a 1-year period from October 1, 2009, to September 30, 2010 (water year 2010) to determine the percentage of time the reach was in gaining-stream or losing-stream conditions. Piezometers at both reaches were installed in vertical nests (different depths at the same location) and horizontal transects (different locations at similar depths) to enable mean daily vertical and horizontal hydraulic gradients to be computed between the river and groundwater (figs. 36C, 37C). If the mean daily horizontal and vertical gradients were negative, the groundwater

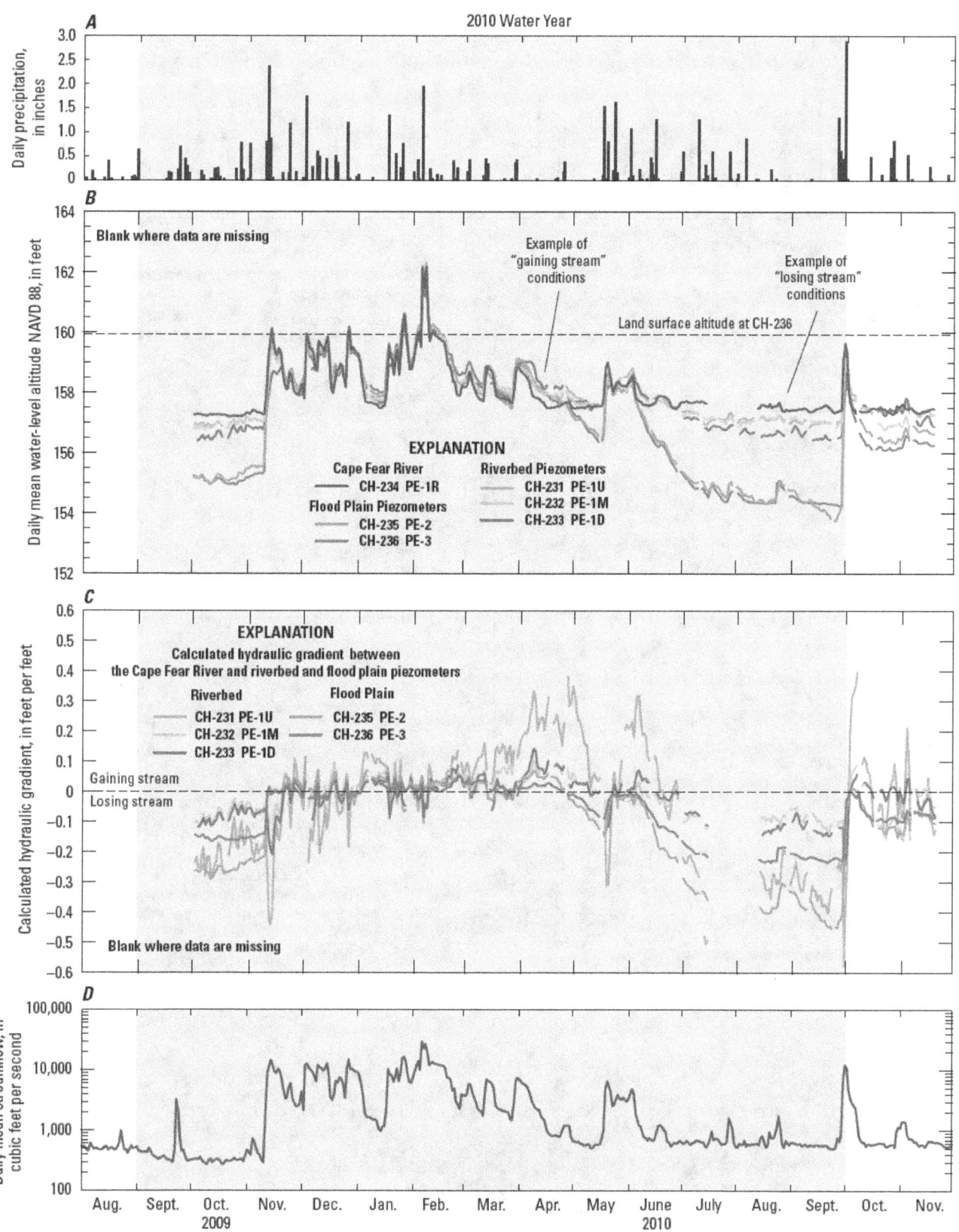

Figure 36. (*A*) Daily precipitation measured at the Siler City Airport (SILR), (*B*) daily mean water level and (*C*) calculated hydraulic gradients at the regional power plant piezometer transect (PZ1) near Corinth, and (*D*) daily mean streamflow measured in the Cape Fear River at Lillington (site 28), North Carolina.

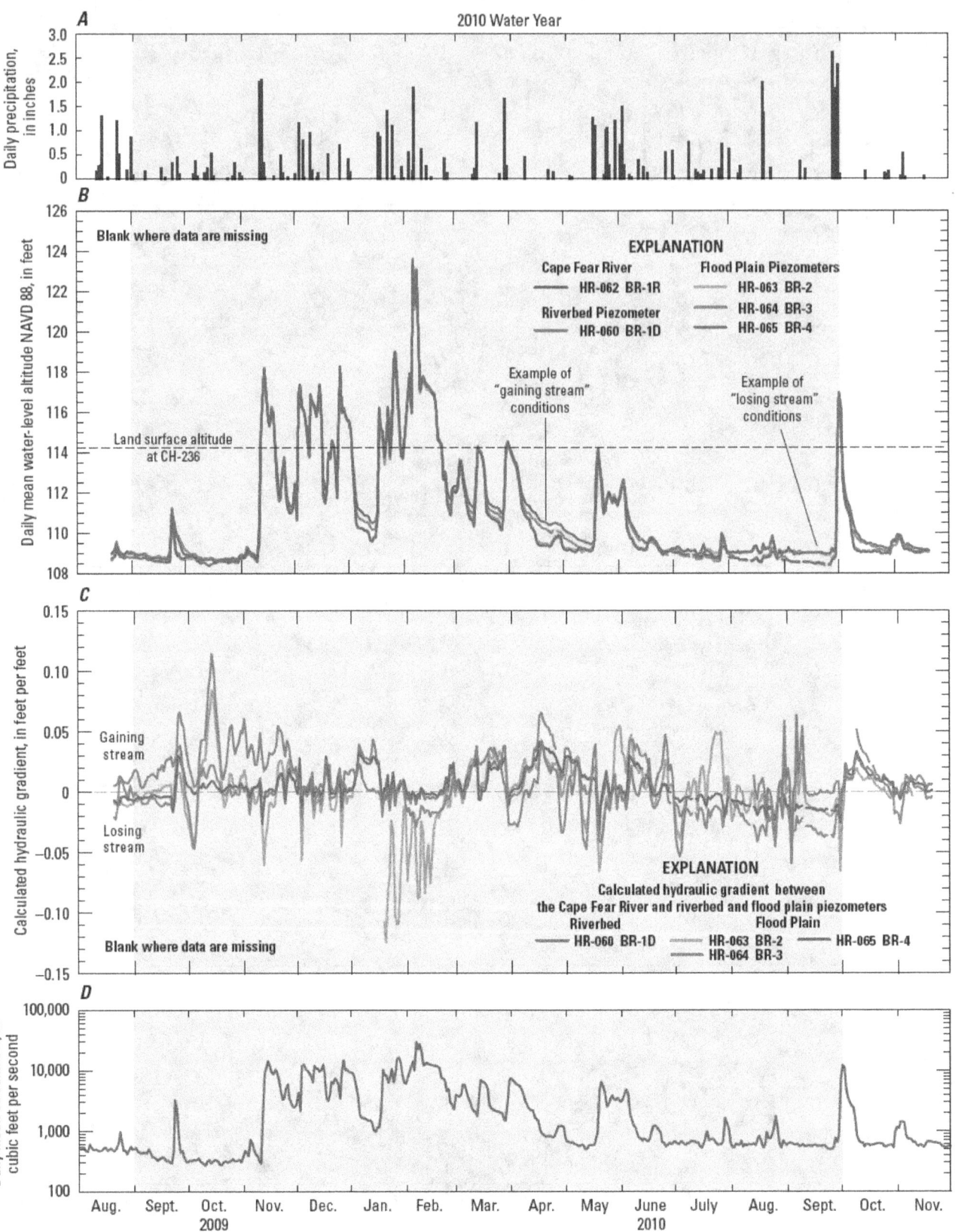

Figure 37. (*A*) Daily precipitation measured at the Fayetteville Public Works Commission (313017) Cooperative station, (*B*) daily mean water level and (*C*) calculated hydraulic gradients at the Bradley Road piezometer transect (PZ2) near Lillington, and (*D*) daily mean streamflow measured in the Cape Fear River at Lillington (site 28), North Carolina.

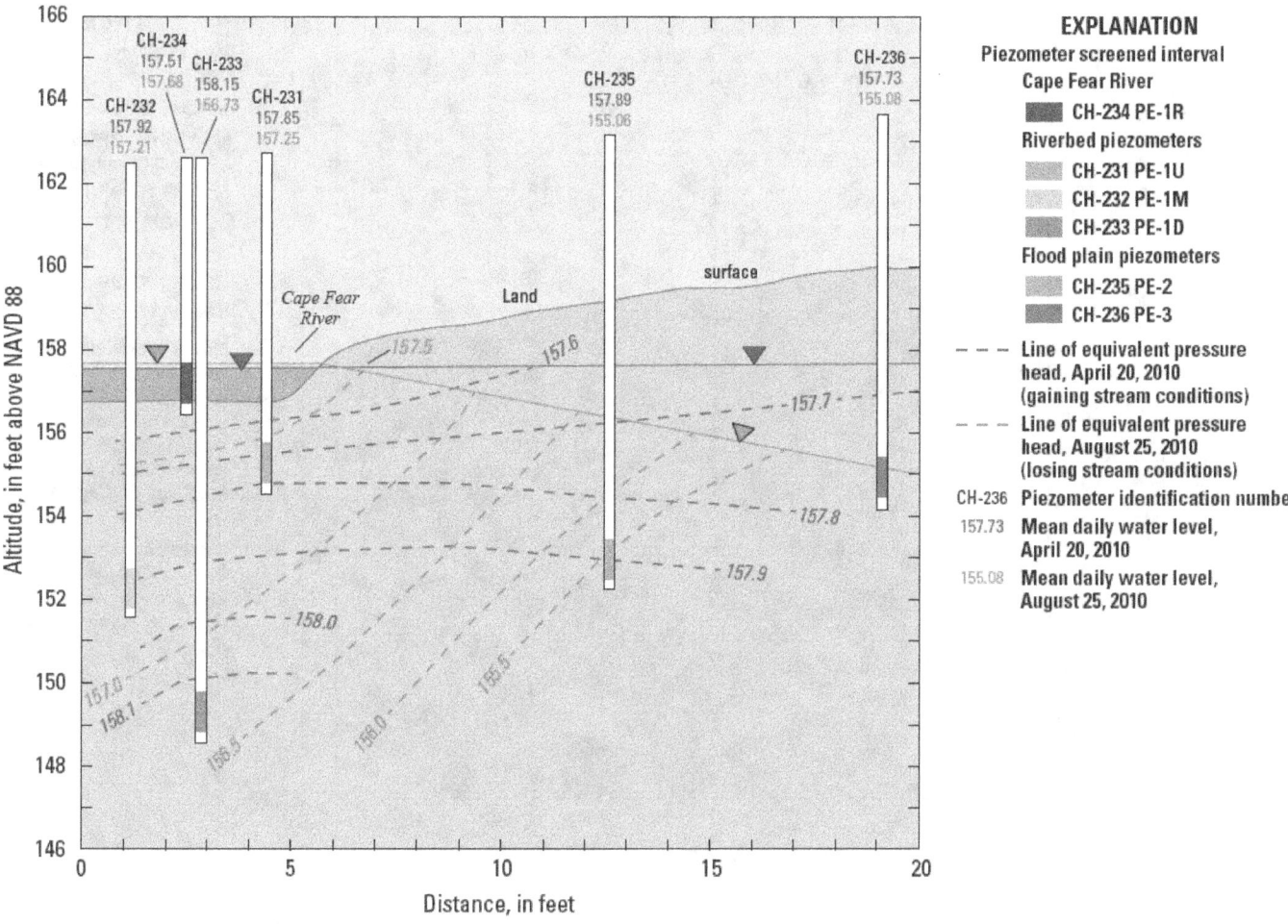

Figure 38.   Piezometers and interpolated lines of equivalent pressure head under gaining- and losing-stream conditions at the regional power plant transect (PZ1) near Corinth, North Carolina (location shown in fig. 1).

flow was downward and outward from the river to the bed and flood-plain sediments, indicating the hydraulic head was less at depth and losing-stream conditions were assigned for the day. Conversely, if the groundwater flow was upward and toward the river from the bed and flood-plain sediments, the mean daily horizontal and vertical gradients were positive, indicating the hydraulic head was greater at depth and gaining-stream conditions were assigned for the day. It was also possible for a reach to be gaining in the vertical direction but losing in the horizontal direction or vice versa. This condition most likely occurs when the vertical or horizontal gradients are reversing because of changes (increase or decrease) in streamflow from impoundment releases or precipitation events but also may occur because the piezometers in the transect intercept different parts of the groundwater-flow path during these events. For example, the vertical flow path beneath the river may have a positive gradient (gaining stream) because of regional groundwater recharge at higher altitudes, yet the horizontal flow path may have a negative gradient (losing stream) because of a rapid rise in river stage.

The category "both" was assigned to this condition (figs. 36B, C, 37B, C). The distribution of hydraulic head under gaining- and losing-stream conditions as measured on April 20 and August 25, 2010, in the regional power plant and Bradley Road reaches, respectively, is displayed in figures 38, 39. Note that the hydraulic gradient under losing-stream conditions on August 25, 2010, in the regional power plant reach is about five times greater than under gaining-stream conditions measured on April 20, 2010 (fig. 38). In contrast, the hydraulic gradient in the Bradley Road reach is similar in magnitude under both gaining- and losing-stream conditions (fig. 39).

Figures 36C and 37C display the distribution of losing- and gaining-stream conditions during the monitoring period. For water year 2010, the Cape Fear River in the regional power plant reach was in a losing-stream condition 38 percent of the year, a gaining-stream condition 26 percent of the year, and both 28 percent of the year. In the Bradley Road reach during the same period, the Cape Fear River was in a losing-stream condition 13 percent of the year, a gaining-stream condition 34 percent of the year, and both 43 percent of the year.

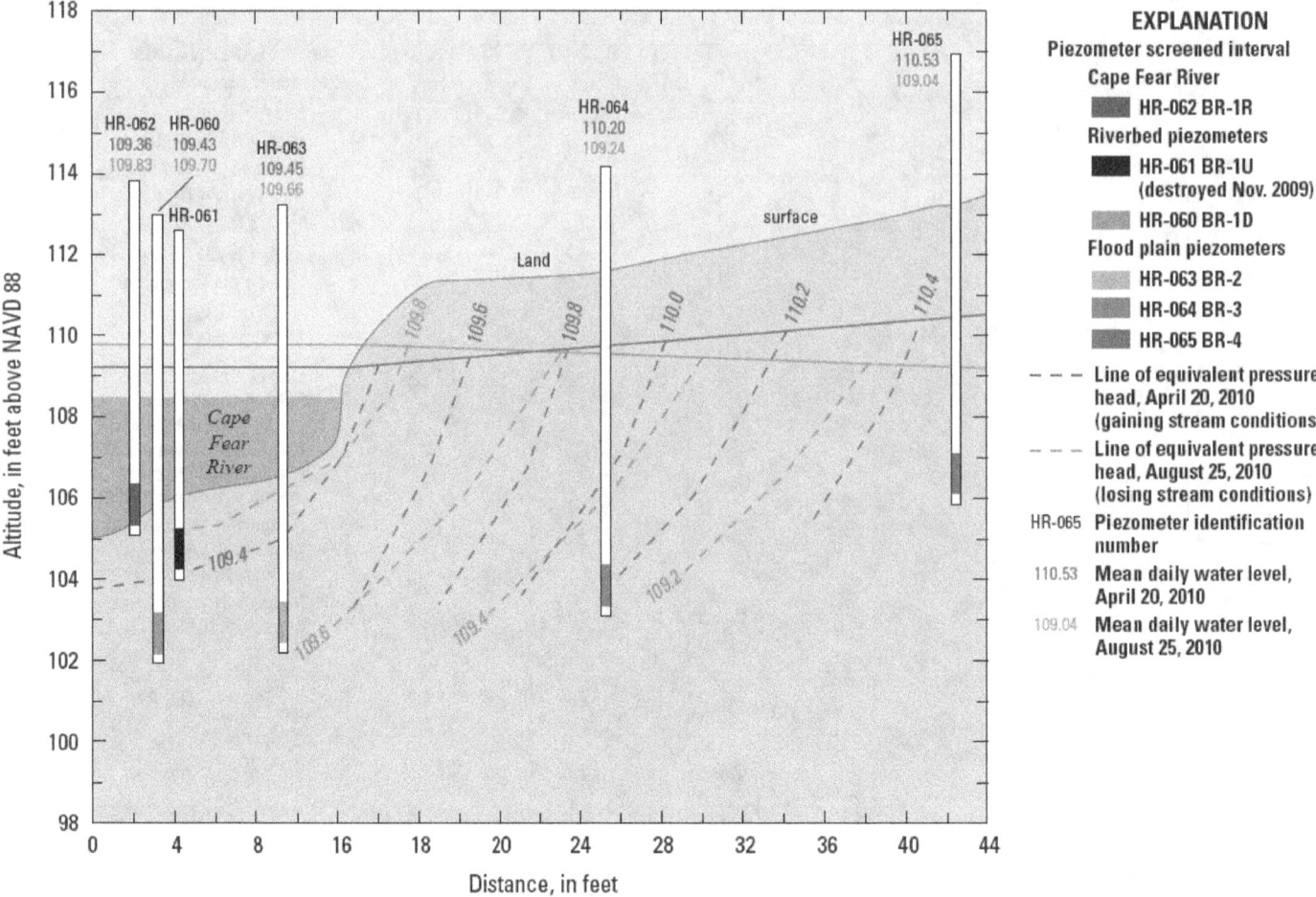

Figure 39. Piezometers and interpolated lines of equivalent pressure head under gaining- and losing-stream conditions at the Bradley Road transect (PZ2) near Lillington, North Carolina (location shown in fig. 1).

Not enough water-level data were available to calculate a gain or loss during part of July and August 2010 and intermittently throughout the study period because of pressure transducer failures in both the regional power plant and Bradley Road reaches, which resulted in a loss of 8 percent and 10 percent of data for the year, respectively. Both vertical and horizontal gradients were frequently two to three times greater under both gaining and losing conditions in the regional power plant reach than those calculated for the Bradley Road reach. This difference is likely because of differences in soils in the reaches (fig. 9). Soils in the regional power plant reach contain more silt and clay, derived mostly from Mesozoic basin sedimentary rocks (Hayes, 2006), which retard the movement of water more than the sandy loam soils, derived from metamorphic rocks of the Cary sequence of the Carolina Terrane (Spangler, 1994) in the Bradley Road reach.

Water-level data collected for more than 1 year at the Progress Energy and Bradley Road reaches indicate that differing hydrologic conditions occur in each reach. During water year 2010, the regional power plant reach lost water through groundwater seepage from the Cape Fear River into the underlying aquifer and adjacent flood plain on about three times as

many days as the Bradley Road reach, with most of the losing days occurring during the summer months when ET processes increased. In contrast, the Bradley Road reach gained water through groundwater discharge from the underlying aquifer and adjacent flood plain on about 30 more days during water year 2010 than in the regional power plant reach, and most of these days occurred during the late spring.

Causes of Flow Losses Based on Results of Data Analyses

Consideration of all data collected and (or) analyzed during this investigation indicates a study reach with complex flow patterns affected by numerous factors and resulting flow losses. Flows often are affected concurrently by several factors, and so the causes of flow loss could not be solely attributed to any one factor as the primary cause. The factors considered in this investigation were (1) computation of discharges at streamgages in the study reach and flow releases from Jordan Lake Dam, (2) effects of diversion and (or) storage at Buckhorn Dam and the Lockville hydropower station, (3) water use associated with industrial and municipal intakes in the study area, (4) losses by evaporation from the study reach, and (5) groundwater and surface-water interaction along the study reach. Analyses of flow data completed for the 1983–2010 water years indicate the median flow loss is 37 ft³/s, based on the assessment of days when steady conditions were in effect (that is, little to no change in flow conditions from the previous day).

Water diversions and evaporative losses were determined to be sufficient on some days (particularly during base-flow periods) to exceed the net gain in flows between the upstream and downstream ends of the study reach. While consumptive use associated with the regional power plant adjacent to the study reach was not fully quantified during the investigation, the net consumptive use associated with five industrial and two municipal intakes was computed to be about 2.5 ft³/s (about 1.6 Mgal/d) and from 23 to 36 ft³/s (35.6 to 55.7 Mgal/d), respectively. Potential flow loss from open-water evaporation in the study reach was estimated to be in the range of 4 to 14 ft³/s during May through October. The summation of these ranges clearly exceeds the median flow loss of 37 ft³/s, indicating flow loss from the study reach due to water diversions and evaporation is real. Ratings analyses completed for the streamgages at Deep River at Moncure (site 3) and Cape Fear River at Lillington (site 28) did not indicate any particular time during the 1982–2011 water years when the use of stage-discharge ratings resulted in computed discharges that indicated false flow losses. Similarly, discharge measurements made at the streamgage on the Haw River downstream from Jordan Dam provided general confirmation of the reported flow releases from Jordan Lake.

Groundwater data collected during 2009–2010 indicate the possibility of localized flow loss during the summer, particularly in the reach above Buckhorn Dam. No indication of unusual patterns was evident, however, that would cause substantial flow loss as a result of groundwater and surface-water interaction at the river bottom. At the regional power plant piezometer reach upstream from Buckhorn Dam, the groundwater level quickly declined below the river level during the base-flow period in late summer and early fall, which indicates losing-stream conditions. Groundwater-level altitudes measured in the flood-plain piezometers were commonly more than 3 ft lower than the Cape Fear River during the summer months. In contrast, groundwater-level altitudes measured in the Bradley Road piezometers were rarely more than 1 ft lower than the river stage, although still representative of losing-stream conditions. During the year of data collection, the Cape Fear River was in a losing-stream condition 38 and 13 percent of the year at the regional power plant and Bradley Road reaches, respectively. The higher percentage at the regional power plant reach is attributed to storage in the impounded reach upstream from Buckhorn Dam.

The distributed temperature survey completed August 2009 on the Cape Fear River at Raven Rock State Park indicated that, under conditions similar to those present during the survey (high ambient temperature, low river stage, high ET processes, and low flow), the study reach is not gaining. In contrast, the aerial TIR survey completed February 27, 2010,

indicated that under some hydrologic conditions during the winter, much of the study reach gained an undetermined quantity of water through groundwater seepage in selected reaches along the main stem.

Summary and Conclusions

During 2008–2010, the U.S. Geological Survey (USGS) conducted a hydrologic investigation in cooperation with the Triangle J Council of Governments Cape Fear River Flow Study Committee and the North Carolina Division of Water Resources to collect hydrologic data to help determine if suspected flow losses in the Cape Fear River between B. Everett Jordan Lake and Lillington, North Carolina, are real or apparent. During some low-flow conditions, the sum of flow releases from Jordan Lake Dam and observed streamflow at a long-term streamgaging station on the Deep River at Moncure in Chatham County is greater than the observed streamflow downstream at the streamgage at Lillington. In addition, flows at the Lillington streamgage have been observed to decrease by 50 to 150 ft³/s over a period of 12 to 24 hours for unknown reasons. Assessment of the flow patterns in the study reach required a multidiscipline approach for surface-water and groundwater data collection and analyses.

Net differences in flows between the upstream and downstream extents of the study reach were determined by summing the daily discharges at Jordan Lake Dam, Deep River at Moncure, and Buckhorn Creek near Corinth and subtracting that sum from the discharge at Cape Fear River at Lillington. Among 10,227 days during the 1983–2010 water years, flow loss occurred on 2,941 days (28.8 percent) based on comparison of the daily discharge between the inputs and the Lillington streamgage. However, flow loss occurred on 408 days (4.0 percent) during conditions that were relatively steady with respect to records for the previous day. The flow loss among these 408 days ranged from 0.49 to 2,150 ft³/s with a median flow loss of 37.2 ft³/s. In terms of flow-loss amounts, analyses of histograms indicated that approximately 64 percent of the flow losses were 50 ft³/s or less. The months with the highest number of days with flow losses were June (16.7 percent), September (16.9 percent), and October (19.4 percent).

Part of the surface-water data collected during the study included discharge measurements made at selected locations along the main stem between Jordan Lake Dam and the Lillington streamgage during a series of six synoptic measurements in 2009. The series of synoptic measurements were conducted on April 27, June 25, July 22, August 18, September 9, and October 1. Several patterns were noticeable from the graphed data of each series of measurements when viewed from a greater perspective. The largest water diversion for use as cooling water at the regional power plant occurs just downstream from the confluence of the Haw and Deep Rivers. Downstream from Buckhorn Dam, little gain or loss was noted between the dam and Raven Rock State Park, although some minor fluctuations in flow patterns appeared to occur between the State park and the Lillington streamgage.

Analyses of discharge measurements and ratings for the two streamgages at Deep River at Moncure and Cape Fear River at Lillington were completed as part of this investigation to address a concern that computed discharge records at these two sites could perhaps be inaccurate, resulting in false flow loss. A total of 99 discharge measurements were made at the Moncure streamgage during the 1982–2011 water years, ranging from 27.9 to 21,500 ft³/s; 66 measured discharges were 500 ft³/s or less. Among these 66 measurements, rating shifts were applied after 45 measurements were completed; 14 shifts were positive, and 31 shifts were negative. A total of 103 discharge measurements were made at the Lillington streamgage during the same period, ranging from 133 to 28,800 ft³/s; 51 measured discharges were 1,000 ft³/s or less. Among these 51 measurements, rating shifts were applied after 11 measurements were completed; 10 shifts were positive, and 1 shift was negative. Examination of the percentage differences (whether shifted or unshifted) at both streamgages likewise did not indicate a particular time during the 1982–2011 water years when the values were consistently positive or negative, and no bias was indicated that could result in falsely computed flow losses.

A comparison was made between 34 measured discharges at a streamgage on the Haw River below B. Everett Jordan Lake near Moncure and reported hourly flow releases from Jordan Lake Dam. The discharge measurements ranged from 94.7 to 9,970 ft³/s, and 25 measurements were less than 2,000 ft³/s. Comparison of these measurements and flow releases indicates the measured discharges ranged from 75 to about 140 percent of the concurrent hourly flow releases among the 34 measurements and within plus or minus 10 percent for 28 measurements, which provides general support of the current discharge computation tables used for reporting Jordan Lake Dam flow releases.

A stage gage was operated during the study on the Cape Fear River at Buckhorn Dam near Corinth to collect continuous stage-only records. Elevation computed from stage records collected during the study ranged from 157.32 ft on October 9 and 12, 2009, to 161.42 ft on February 6, 2010. Flow over the dam was observed along its length during the study, and the above range in stage records indicates that flow loss in the study reach is not attributed to river-level fluctuations at the dam.

Water-use information and (or) data were obtained for five industrial facilities, a regional power plant, two municipalities, one small hydropower facility on the Deep River, and one quarry operation adjacent to the Deep River. The largest water users are the regional power plant, the small hydropower facility, and the two municipalities.

Water-use data collected at the five industrial facilities during the study indicate the maximum water consumption among these facilities is about 2.5 ft³/s (about 1.6 Mgal/d). Combining the range of withdrawals for the two municipalities indicates that from 23 to 36 ft³/s (35.6 to 55.7 Mgal/d) was removed from the study reach during the winter and summer periods. When adding the maximum water consumption for all five industries and quarry operation, the total water-use diversions ranged from almost 25.5 to 38.5 ft³/s (39.5 to 59.5 Mgal/d) during the winter and summer periods. This range in water-use diversions is equivalent to 69 to 104 percent of the 37 ft³/s median flow loss.

The Lockville hydropower station is on the Deep River about 1 mi downstream from the streamgage near Moncure. Under the current (2011) operation of one turbine, the discharge diverted through the canal between the dam and powerhouse during maximum power generation is between 500 and 600 ft³/s. Information obtained during two visits to the facility indicated that run-of-river operations did not appear to be a major factor behind flow losses in the study reach.

The largest water user in the study area is a regional power producer at a coal-fired power generation plant located immediately adjacent to the Cape Fear River just downstream from the confluence of the Haw and Deep Rivers. Data describing daily water withdrawals were supplied by the regional power producer and compared to discharge records at a USGS streamgage on the diversion canal downstream from the power plant. Estimated differences between the intake and canal discharge data (or consumptive losses) typically ranged from 10 to 40 ft³/s (6.5 to about 25.8 Mgal/d) with amounts exceeding 50 ft³/s (32.3 Mgal/d) on some days. A large part of this range is higher than the approximately 13.4 ft³/s (8.65 Mgal/d) estimate in water loss from the cooling towers provided from engineers with the regional power producer. Uncertainty surrounding reasonable estimates of consumption remained in effect at the end of the study.

Data concerning evaporative losses were compiled by using two approaches: (1) analysis of available pan-evaporation data from a collection station nearby in Chapel Hill, N.C., and (2) compilation of reference open-water evaporation data computed by the State Climate Office of North Carolina using the Penman-Monteith method. Comparison of estimated evaporative losses indicated substantial differences between the two methods of computing evaporation, particularly during the summer months. Both methods were considered by estimating the potential flow loss by evaporation from the main stem and the Deep River to be in the range of 4 to 14 ft³/s during May through October, equivalent to 10 to 38 percent of the 37 ft³/s median flow loss.

Daily water-use diversions and evaporation losses were compared to flow-loss occurrences during the period April 2008 to September 2010. The comparison was completed to assess how the magnitude of total losses from water consumption and evaporation compared to calculated flow losses. During the period of interest, flow analyses indicated 289 days of flow loss occurrences at the streamgage on the Cape Fear River at Lillington, with filtered flow losses occurring on 34 days during the same period. Among these 34 days, the total consumptive water-use diversions for the five industries and two municipalities exceeded the computed flow losses on 26 days. With evaporative losses added, the total losses exceeded the computed flow losses on 2 additional days.

During the latter stages of the investigation, a connection was established between the occasional sudden changes in flow at the streamgage on the diversion canal adjacent to the Cape Fear River and downstream flows at the Lillington streamgage, resulting in an apparent flow loss. An assessment of the daily discharges and subsequent inspection of the instantaneous-value hydrographs for the canal streamgage indicated at least 24 instances from the start of data collection in April 2008 through December 2010 when the flows inexplicably decreased or increased by 100 to more than 200 ft^3/s. For 10 of these 24 instances, flows at the Lillington streamgage were at or below 600 ft^3/s. The instantaneous hydrographs showed a fairly consistent cause-and-effect relation. Inspection of the hydrographs indicated that, when discharge through the canal decreased or stopped for a short period, that amount of discharge was immediately available for travel in the Cape Fear River, which resulted in increased discharge at the downstream Lillington streamgage beginning about 8 hours later. When discharge from the Cape Fear River was diverted into the canal, there was a time-lagged decrease in discharges downstream at the Lillington streamgage beginning anywhere between 8 and 16 hours later.

Several complementary methods were applied to assess conditions of gain or loss through groundwater and surface-water interaction within regional (large scale) and local (small scale) confines of the Cape Fear River. A geophysical survey using the GEM–2 digital multifrequency electromagnetic-induction instrument was completed on the main stem during 2009. The geophysical survey was used to identify possible locations of magnetic anomalies indicating the presence of diabase dikes (altering groundwater discharge from or recharge to a river), thereby enabling increased focus on reaches where additional groundwater data, in particular the distributed temperature sensing survey, could be collected. Inspection of mapped anomalies from the study-area scale revealed that most occurrences of increased magnetic anomalies were detected downstream from Buckhorn Dam in the Carolina terrane, Cary sequence rocks and relatively few occurrences were detected upstream.

About 3,000 ft of fiber-optic cable was deployed on the bed of the Cape Fear River at the Raven Rock State Park reach August 11–13, 2009, to determine if the presence of diabase dikes could be preferentially directing groundwater discharge. The average water depth was 1.84 ft and ranged from 0.48 to 3.83 ft. Over the 40 hours of temperature monitoring with the fiber-optic distributed temperature sensor (FO-DTS), water temperatures were dominated by the diurnal effects of solar heating and night-time cooling. No temperature anomalies of colder water were measured during the survey, which indicated that at the time of the survey, the Cape Fear River was a no-flow or losing stream in that particular river reach.

An aerial thermal-infrared survey was conducted over the Haw and Cape Fear Rivers on February 27, 2010, from Jordan Lake Dam to Lillington to qualitatively delineate areas of preferential discharge on the basis of the contrast between warm groundwater discharge and cold surface-water temperatures. Discharge generally was noted as diffuse seepage, but in a few cases springs were detected as inflow as a discrete point of discharge.

Two reaches of the Cape Fear River were selected for groundwater monitoring, each with a transect of piezometers installed in the flood plain (regional power plant and Bradley Road). Water-level altitudes at the regional power plant and Bradley Road reaches were analyzed over a 1-year period from October 1, 2009, to September 30, 2010 (water year 2010). In the regional power plant reach, the Cape Fear River was in a losing-stream condition 38 percent of the year, a gaining-stream condition 26 percent of the year, and both conditions 28 percent of the year. In the Bradley Road reach during the same period, the Cape Fear River was in a losing-stream condition 13 percent of the year, a gaining-stream condition 34 percent of the year, and both conditions 43 percent of the year. Groundwater data collected as part of this study represent only a brief period of time and may not represent conditions present at all times and for all years. However, the data indicated that during the winter, the study area gained an undetermined quantity of water through seepage.

Consideration of all data collected and (or) analyzed during this investigation indicates a study reach with complex flow patterns affected by numerous factors and resulting flow losses. Flows often are affected concurrently by several factors, and the causes of flow loss could not be solely attributed to any one factor as the primary cause. Water diversions and evaporative losses were determined to be sufficient on some days (particularly during base-flow periods) to exceed the net gain in flows between the upstream and downstream ends of the study reach. Losses due to diversions and evaporation can exceed the median flow loss of 37 ft^3/s, indicating flow loss from the study reach is real. Groundwater data collected during 2009–2010 indicate the possibility of localized flow loss during the summer, particularly in the impounded reach above Buckhorn Dam. No indication of unusual patterns was evident, however, that would cause substantial flow loss as a result of groundwater and surface-water interaction at the river bottom.

Selected References

Allen, R.G., Pereira, L.S., Raes, Dirk, and Smith, Martin, 1998, Crop evapotranspiration—Guidelines for computing crop water requirements: Rome, Italy, Food and Agriculture Organization (FAO) of the United Nations, FAO Irrigation and Drainage Paper 56, accessed in November 2011 at *http://www.fao.org/docrep/X0490E/x0490E00.htm*.

Bales, J.D., Chapman, M.J., Oblinger, C.J., and Robbins, J.C., 2003, North Carolina District Science Plan, U.S. Geological Survey Open-File Report 2004–1025, 31 p., accessed on November 7, 2011, at *http://nc.water.usgs.gov/reports/ofr041025/pdf/report.pdf*.

Banks, W.S.L., Paylor, R.L., and Hughes, W.B., 1996, Using thermal-infrared imagery to delineate ground-water discharge: Ground Water, v. 34, no. 3, p. 434–443.

Belknap, W., and Naiman, R.J., 1998, A GIS and TIR procedure to detect and map wall-base channels in western Washington: Journal of Environmental Management, v. 52, p. 147–160.

Blodgett, J.C., Walters, J.R., and Borchers, J.W., 1992, Streamflow gains and losses and selected flow characteristic of Cottonwood Creek, North-Central California, 1982–85: U.S. Geological Survey Water-Resources Investigations Report 92–4009, 19 p.

Bossong, C.R., Caine, J.S., Stannard, D.I., Flynn, J.L., Stevens, M.R., and Heiny-Dash, J.S., 2003, Hydrologic conditions and assessment of water resources in the Turkey Creek watershed, Jefferson County, Colorado, 1998–2001: U.S. Geological Survey Water-Resources Investigations Report 03–4034, 140 p.

Burt, E.R., Carpenter, P.A., III, McDaniel, R.D., and Wilson, W.F., 1978, Diabase dikes of the eastern Piedmont of North Carolina: North Carolina Geological Survey Information Circular 23, 12 p., 1 pl.

Daniels, R.B., Buol, S.W., Kleiss, H.J., and Ditzler, C.A., 1999, Soil systems in North Carolina: North Carolina State University, Department of Soil Science, Technical Bulletin 314, 118 p.

Dumouchelle, D.H., 2008, Streamflow gains and losses for Hellbranch Run, Franklin County, Ohio, August 2007: U.S. Geological Survey Scientific Investigations Report 2008–5191, 7 p.

Executive Office of the President of the United States, 2007, A strategy for federal science and technology to support water availability and quality in the United States: National Science and Technology Council, Committee on Environment and Natural Resources, Subcommittee on Water Availability and Quality, 35 p.

Farnsworth, R.K., and Thompson, E.S., 1982, Mean monthly, seasonal, and annual pan evaporation for the United States: National Oceanic and Atmospheric Administration, NOAA Technical Report NWS 34, 91 p.

Fishman, M.J., ed., 1993, Methods of analysis by the U.S. Geological Survey National Water Quality Laboratory—Determination of inorganic and organic constituents in water and fluvial sediments: U.S. Geological Survey Open-File Report 93–125, 217 p.

Freeman, L.A., Carpenter, M.C., Rosenberry, D.O., Rousseau, J.P., Unger, Randy, and McLean, J.S., 2004, National field manual for the collection of hydrologic data—Use of submersible pressure transducers in water-resources investigations: U.S. Geological Survey Techniques of Water-Resources Investigations, book 8, chap. A3 [variously paged].

Geophex, 2011, GEM 2 Data examples, accessed March 4, 2011, at *http://www.geophex.com/GEM2/Data%20 examples%20GEM–2/data_examples.htm*.

Hayes, R.D., 2006, Soil survey of Chatham County, North Carolina: U.S. Department of Agriculture, Natural Resources Conservation Service, 673 p.

Healy, R.W., Winter, T.C., LaBaugh, J.W., and Franke, O.L., 2007, Water budgets—Foundations for effective water-resources and environmental management: U.S. Geological Survey Circular 1308, 90 p.

Hibbard, J.P., Stoddard, E.F., Secor, D.T., and Dennis, A.J., 2002, The Carolina zone—Overview of Neoproterozoic to early Paleozoic peri-Gondwanan terranes along the eastern flank of the southern Appalachians: Earth-Science Reviews, v. 57, p. 299–339.

Hibbard, J.P., van Staal, C.R., Rankin, D.W., and Williams, H., 2006: Lithotectonic map of the Appalachian Orogen, Canada-United States of America: Geological Survey of Canada, Map 2096A, scale 1:1,500,000, 1 pl.

Lee, K.K., and Risley, J.C., 2001, Estimates of ground-water recharge, base flow, and stream reach gains and losses in the Willamette River Basin, Oregon: U.S. Geological Survey Water-Resources Investigations Report 01–4215, 52 p., 1 pl.

LeGrand, H.E., 2004, A master conceptual model for hydrogeological site characterization in the Piedmont and Mountain region of North Carolina, A guidance manual: Raleigh, North Carolina Department of Environment and Natural Resources, Division of Water Quality, Groundwater Section, 55 p.

McSwain, K.B., Bolich, R.E., Chapman, M.J., and Huffman, B.A., 2009, Water-resources data and hydrogeologic setting at the Raleigh hydrogeologic research station, Wake County, North Carolina, 2005–2007: U.S. Geological Survey Open-File Report 2008–1377, 48 p.

Moix, M.W., and Galloway, J.M., 2004, Base flow, water quality, and streamflow gain and loss of the Buffalo River, Arkansas, and selected tributaries, July and August 2003: U.S. Geological Survey Scientific Investigations Report 2004–5274, 36 p.

Monteith, J.L., 1965, Evaporation and environment, *in* Fogg, G.E., ed.,The state and movement of water in living organisms, Symposium of the Society for Experimental Biology, v. 19: New York, N.Y., Academic Press, Inc., p. 205–234.

Mueller, D.S., and Wagner, C.R., 2009, Measuring discharge with acoustic Doppler current profilers from a moving boat: U.S. Geological Survey Techniques and Methods 3A–22, 72 p., accessed in July 2011 at *http://pubs.usgs.gov/tm/3a22/*.

North Carolina Department of Environment and Natural Resources, Division of Land Resources, Dam Safety Program, 2010, North Carolina Dam Inventory, accessed June 30, 2011, at *http://www.dlr.enr.state.nc.us/pages/ damsafetyprogram.html*.

North Carolina Department of Environment and Natural Resources, Division of Water Resources, 2007, Water withdrawal and transfer registration, 2007 annual water use report, prepared for R.D. Lee Farms, Inc.: accessed May 17, 2010, at *http://www.ncwater.org/Permits_and_Registration/ Water_Withdrawal_and_Transfer_Registration/report/ view/0711–0001/2007.*

North Carolina Department of Environment and Natural Resources, Division of Water Resources, 2009, Water withdrawal and transfer registration, 2009 annual water use report, prepared for Wakestone Corporation Moncure quarry: accessed May 17, 2010, at *http://www.ncwater. org/Permits_and_Registration/Water_Withdrawal_and_ Transfer_Registration/report/view/0202-0001/2009.*

North Carolina Geological Survey, 1985, Geologic map of North Carolina: Raleigh, North Carolina Geological Survey, scale 1:500,000.

Penman, H.L., 1948, Natural evaporation from open water, bare soil, and grass: Proceedings of the Royal Society of London, A193, p. 120–146.

Piper, A.M., 1953, A graphic procedure in the geochemical interpretation of water analyses: U.S. Geological Survey, Ground-Water Chemistry Notes, no. 12, 14 p.

Ragland, B.C., Barker, R.G., and Robinson, J.B., 2003, Water resources data, North Carolina, water year 2002, v. 1B, surface-water records: U.S. Geological Survey Water-Data Report, NC–02–1B, 645 p.

Rantz, S.E., and others, 1982, Measurement and computation of streamflow, v. 1. Measurement of stage and discharge: U.S Geological Survey Water-Supply Paper 2175, p. 1–284.

Rantz, S.E., and others, 1982, Measurement and computation of streamflow, v. 2. Computation of discharge: U.S. Geological Survey Water-Supply Paper 2175, p. 285–631.

Reinemund, J.A., 1955, Geology of Deep River coal field North Carolina: U.S. Geological Survey Professional Paper 246, 159 p., 8 pls.

Rosenberry, D.O., and LaBaugh, J.W., eds., 2008, Field techniques for estimating water fluxes between surface water and ground water: U.S. Geological Survey Techniques and Methods 4–D2, 128 p.

Sauer, V.B., and Meyer, R.W., 1992, Determination of error in individual discharge measurements: U.S. Geological Survey Open-File Report 92–144, 21 p. , accessed in July 2011 at *http://pubs.usgs.gov/of/1992/ofr92-144/.*

Selker, J.S., Thévenaz, Luc, Huwald, Hendrik, Mallet, Alfred, Luxumberg, Wim, van de Giesen, Nick, Stejskal, Martin, Zeman, Josef, Westhoff, Martijn, and Parlange, M.B., 2006, Distributed fiber-optic temperature sensing for hydrologic systems: Water Resources Research, v. 42, 8 p.

Spangler, D.G., 1994, Soil survey of Harnett County, North Carolina: U.S. Department of Agriculture, Natural Resources Conservation Service, 171 p.

State Climate Office of North Carolina, 2010, FAO56 Penman-Monteith Reference Evapotranspiration Estimates: North Carolina State University, accessed November 7, 2011, at *http://www.nc-climate.ncsu.edu/et.*

Stonestrom, D.A., and Constantz, Jim, eds., 2003, Heat as a tool for studying the movement of ground water near streams: U.S. Geological Survey Circular 1260, 96 p.

Torgersen, C.E., Faux, R.N., McIntosh, B.A, Poage, N.J., and Norton, D.J., 2001, Airborne thermal remote sensing for water temperature assessments in rivers and streams: Remote Sensing of Environment, v. 76, issue 3, p. 386–398.

Turco, M.J., East, J.W., and Milburn, M.S., 2007, Base flow (1966–2005) and streamflow gain and loss (2006) of the Brazos River, McLennan County to Fort Bend County, Texas: U.S. Geological Survey Scientific Investigations Report 2007–5286, 27 p.

Turnipseed, D.P., and Sauer, V.B., 2010, Discharge measurements at gaging stations: U.S. Geological Survey Techniques and Methods book 3, chap. A8, 87 p. , accessed in July 2011 at *http://pubs.usgs.gov/tm/tm3-a8/.*

U.S. Army Corps of Engineers, 1993, Installing monitoring wells/piezometers in wetlands: Wetlands Research Program Technical Note HY-IA–3.1, 14 p., accessed June 24, 2011, at *http://el.erdc.usace.army.mil/elpubs/pdf/hyia3-1.pdf.*

U.S. Environmental Protection Agency, Multi-Resolution Land Characteristics Consortium, 2007, National Land Cover Data classification schemes (Level II) and definitions, accessed June 23, 2011, at *http://www.epa.gov/mrlc/ classification.html and http://www.epa.gov/mrlc/definitions. html.*

U.S. Geological Survey, 2001, USGS water data for the Nation: U.S. Geological Survey National Water Information System Web interface, accessed in June 2011 at *http:// waterdata.usgs.gov/nwis/.*

U.S. Geological Survey, 2002, Concepts for national assessment of water availability and use: U.S. Geological Survey Circular 1223, 34 p.

U.S. Geological Survey, 2006, Collection of water samples (ver. 2.0): U.S. Geological Survey Techniques of Water-Resources Investigations, book 9, chap. A4, September 2006, accessed February 22, 2011, at *http://pubs.water.usgs. gov/twri9A4/.*

U.S. Geological Survey, 2007a, Facing tomorrow's challenges—U.S. Geological Survey science in the decade 2007–2017: U.S. Geological Survey Circular 1309, 67 p.

U.S. Geological Survey, 2007b, Water-resources data for the United States, Water Year 2006: U.S. Geological Survey Water-Data Re*port WDR-US–2006, accessed February 26, 2008, at http://pubs.water.usgs.gov/wdr2006.*

U.S. Geological Survey, 2010a, The Cooperative Water Program – Program Priorities for 2011; accessed on November 7, 2011, at *http://water.usgs.gov/coop/priorities.html.*

U.S. Geological Survey, 2010b, North Carolina District operating procedures and guidelines, Chapter C – Quality-assurance plan for surface-water activities of the North Carolina Water Science Center, accessed September 1, 2011, at *http://nc.water.usgs.gov/usgs/info/qaplan/surface.html.*

U.S. Geological Survey, 2011a, Water-resources data for the United States, Water Year 2010: U.S. Geological Survey Water-Data Report WDR-US–2010, accessed November 21, 2011, at *http://wdr.water.usgs.gov/wy2010/search.jsp.*

U.S. Geological Survey, 2011b, The water cycle: accessed March 14, 2012, at *http://ga.water.usgs.gov/edu/watercycle.html.*

Veenhuis, J.E., 2002, Summary of flow loss between selected cross sections on the Rio Grande in and near Albuquerque, New Mexico: U.S. Geological Survey Water-Resources Investigations Report 02–4131, 30 p.

Walters, D.A., Robinson, J.B., and Barker, R.G., 2006, Water resources data, North Carolina, water year 2005, v. 1, surface-water records: U.S. Geological Survey Water-Data Report, NC–05–1, 1,090 p.

Weaver, J.C., and Pope, B.F., 2001, Low-flow characteristics and discharge profiles for selected streams in the Cape Fear River basin, North Carolina, through 1998: U.S. Geological Survey Water-Resources Investigations Report 01–4094, 140 p.

Winter, T.C., Harvey, J.W., Franke, O.L., and Alley, W.M., 1998, Ground water and surface water a single resource: U.S. Geological Survey Circular 1139, 79 p.

Yonts W.L., Giese, G.L., and Hubbard, E.F., 1973, Evaporation from Lake Michie, North Carolina, 1961–**7**1: U.S. Geological Survey Water-Resources Investigations Report 38–73, 27 p.

Appendix 1.   A Brief Primer on Discharge Measurements and Ratings Analyses

In operating a streamgage, records of gage height (or stage) are collected and applied to a stage-discharge rating to determine the discharge associated with a given stage. The stage-discharge rating relates the stage and discharge values for a specific site. The rating is established over time with the collection of discharge measurements made at the streamgage over the range of stages expected at the site. Because of ever-changing conditions in the channel, ratings are dynamic and typically evolve over time as additional discharge measurements are made and used to further update the relation between stage and discharge. For example, channel scour resulting from a large flood event may be sufficient to warrant a revised rating for a streamgage, while flow events of lesser magnitude may result in a temporary "shift" of the rating (fig. 1–1).

Discharge records at USGS streamgages commonly are based on the use of a stage-discharge relation (or rating), which defines the relation between the stage and discharge for a specific location on a stream. The establishment and maintenance of a stage-discharge rating for a streamgage is an on-going process requiring the continual collection and evaluation of discharge measurements over time. As discharge

measurements are collected at a streamgage, the rating is developed and reanalyzed to determine if the current rating is appropriate or if a temporary "rating shift" is needed. A number of factors govern the strength and stability of a rating, one of the most notable being the materials that make up the channel bottom at the streamgage. The channel bottom material at the Moncure and Lillington streamgages is characterized generally by rocks of varying sizes ranging from small cobbles to large boulders. The process of scour and fill under the range of discharges that can occur on a stream may result in a temporary rating shift following a moderate- to high-flow event. Temporary rating shifts also may be required when debris is lodged in the channel following storm events, aquatic growth in the channel, or beaver-dam activity. If the temporary shift remains in effect over the course of time after additional discharge measurements have been obtained, then a new stage-discharge rating may be established for the streamgage.

When a discharge measurement is made at a streamgage, the discharge is computed and three initial steps are completed by the hydrographer. The first is to establish the quality of the measurement by using one of four assessments (excellent,

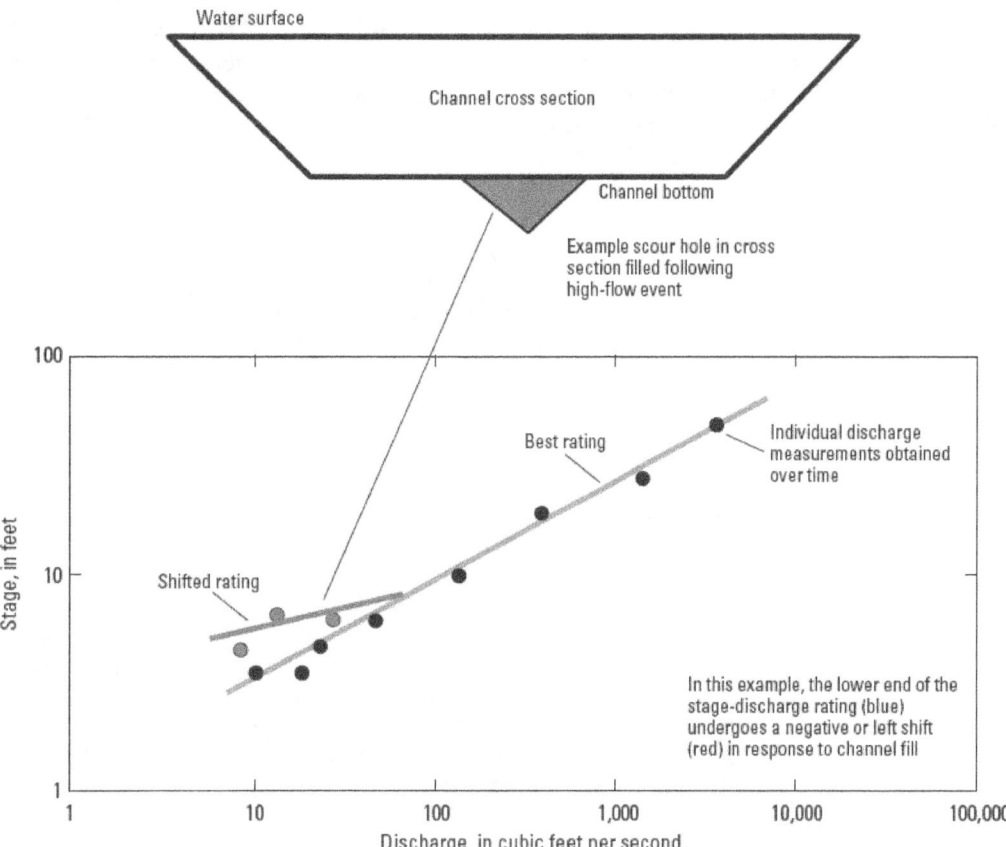

Figure 1–1. Schematic showing theoretical shift in stage-discharge rating in response to changes in river channel geometry.

good, fair, or poor). For measurements made using mechanical velocity meters, the quality is based on the number of sections, percentage of discharge within each section, the mean velocity, and occurrence of angle coefficients (U.S. Geological Survey, 2010b). For discharge measurements made using acoustic meters, the measurement quality is based on the calculated 95-percent confidence interval of base measurement uncertainty (in percent). For all measurements, the condition of equipment, field measuring conditions, and the hydrographer's judgment also factor into the quality of the measurement.

If the base uncertainty for an acoustic measurement is less than or equal to 5 percent, the measurement rating is good (U.S. Geological Survey, 2010b). If the base uncertainty is between 5 and 8 percent, the measurement rating is fair. If the base uncertainty is greater than 8 percent, the measurement rating is poor. Although the assessment is somewhat subjective, the rating of a measurement is an important step and reflects the hydrographer's characterization of the overall quality of the measurement. The measurement quality rating is useful in determining whether any temporary shifts need to be applied to the stage-discharge rating in effect for the streamgage, which affects the discharge records that are computed on the basis of the rating.

In the modern environment of real-time streamflow dissemination, current streamflow data are determined using a base rating curve and a temporary rating shift (if needed and where applicable) as indicated by the most recent discharge measurement(s). This is a critical point of consideration when assessing the accuracy of discharge data continuously being provided through real-time capabilities. Until the next discharge measurement is made at a gaging station, any temporary shift applied to a rating remains in effect until the next measurement can be processed and reviewed to determine if a shift is still necessary. Real-time discharge data, therefore, are considered provisional and subject to revision pending subsequent analysis as additional discharge measurements become available. Because provisional discharge data can be subject to an unknown percentage difference from the base rating, the possibility exists that incorrect assessments of flow loss can occur until the next measurement is made.

The second step completed by the hydrographer following a discharge measurement is to compute an unshifted percentage difference between the measured discharge and base rating discharge. This value is computed by subtracting the base rating discharge from the measured discharge, dividing by the base rating discharge, and expressing the result as a percentage. A positive unshifted percentage difference indicates the measured discharge is *higher* than the base rating discharge, and a negative unshifted percentage difference indicates the measured discharge is *lower* than the base rating discharge.

Based on the quality assigned to the discharge measurement, the magnitude of percentage difference, and other factors observed during the field inspections, the hydrographer may apply a temporary shift to align the base rating with observed conditions at the time of the measurement. Although rare, the hydrographer may elect not to use the measurement to apply any shift in the rating and resulting computation of discharge records.

A shift is measured in stage units (feet) and corresponds to the equivalent change in stage necessary to bring the base rating discharge in sync with the measured discharge, resulting in a shift-adjusted rating. Following the shift, a shifted percentage difference between the measured discharge and the shift-adjusted-rating discharge is computed using the approach described above for the unshifted percentage difference.

More information documenting the computation of discharge at USGS streamgaging stations is available in Rantz and others (1982, v. 1, 2).

Manuscript approved October 2, 2012

For more information about this publication, contact:

Director
USGS North Carolina Water Science Center
3916 Sunset Ridge Road
Raleigh, North Carolina 27607
phone: (919) 571-4000
fax: (919) 571-4041

Or visit the North Carolina Water Science Center
Web site at: http://nc.water.usgs.gov

Prepared by the Raleigh Publishing Service Center

A PDF version of this publication is available online at
http://pubs.usgs.gov/sir/2012/5226